Palgrave Studies in Media and Environmental Communication

Series Editors
Anders Hansen
School of Media, Communication and Sociology
University of Leicester
Leicester, UK

Steve Depoe
McMicken College of Arts and Sciences
University of Cincinnati
Cincinnati, OH, USA

Drawing on both leading and emerging scholars of environmental communication, the Palgrave Studies in Media and Environmental Communication Series features books on the key roles of media and communication processes in relation to a broad range of global as well as national/local environmental issues, crises and disasters. Characteristic of the cross-disciplinary nature of environmental communication, the books showcase a broad variety of theories, methods and perspectives for the study of media and communication processes regarding the environment. Common to these is the endeavour to describe, analyse, understand and explain the centrality of media and communication processes to public and political action on the environment.

More information about this series at
http://www.palgrave.com/gp/series/14612

Tim Jensen

Ecologies of Guilt in Environmental Rhetorics

palgrave
macmillan

Tim Jensen
School of Writing, Literature, and Film
Oregon State University
Corvallis, OR, USA

Palgrave Studies in Media and Environmental Communication
ISBN 978-3-030-05650-6 ISBN 978-3-030-05651-3 (eBook)
https://doi.org/10.1007/978-3-030-05651-3

Cover image: Paulo Oliveira/Alamy Stock Photo
Cover design by eStudio Calamar

This Palgrave Pivot imprint is published by the registered company Springer Nature Switzerland AG
The registered company address is: Gewerbestrasse 11, 6330 Cham, Switzerland

Acknowledgements

Writing a book about gratitude probably would have been more pleasant. This was an emotionally taxing work, which makes the support I received from others all the more meaningful.

Thanks to the editorial team at Palgrave for their enduring support and patience. The book was improved by the insightful comments of reviewers at each stage, for which I am grateful. Special thanks to Bridie McGreavy and Caroline Gottshalk Druschke, who I met at an RSA Institute years ago and whose work has been an inspiration and guide ever since.

This project took shape while on fellowship at the Center for the Humanities at Oregon State University, which provided a refuge to read and think. Writing at Shotpouch Cabin was instrumental to the manuscript's development—thank you to Carly Lettero and the Spring Creek Project for the residency fellowships.

My colleagues in the School of Writing, Literature, and Film are suspiciously good people and unnervingly kindhearted. Special thanks to Lily Sheehan, Megan Ward, Ray Malewitz, Rebecca Olson, George Estreich, Tara Williams, and Peter Betjemann for your mentorship. I'm grateful for the energy emanating from Elena Passarello, Jen Richter, Marjorie Sandor, Keith Scribner, Sue Rodgers, Karen Holmberg, and Kristin Griffin. Boundless thanks to Felicia Phillips. I have deep affection and appreciation for my rhetoric and composition cohort—Lisa Ede, Chris Anderson, Vicki Tolar Burton, Anita Helle, Ana Ribiero, and Kristy Kelly. Ehren Pflugfelder deserves unique thanks, as he has been, at every step and pedal of the way, an impossibly great colleague and friend. Many thanks to Anne-Marie Deitering, Hannah Rempel, Dennis Bennett, Clare Braun, and Wes Snyder for their support.

This book was positively influenced by conversations with many undergraduate and graduate students. Crisp high-fives to Paisley Green, Jessie Heine, jenna, Addy Koneval, Ian Ferris, Brooke Bishop, and Laura Bennett. Marie Wiley's wild wisdom runs throughout these pages. Climb Mount Fuji, O snail, but slowly, slowly.

For their support from day one, thanks to Michael Harker, Aaron McKain, Ben McCorkle, Scott DeWitt, Kay Halasek, Wendy Hesford, and Cindy and Dickie Selfe. My gratitude for Katie Comer runs too deep for words. You'll just have to trust me.

To craftsmen of the good life—Ted Leeson, John Larison, Tim Cleary, Mike Pakula, and Cory Catignani—thanks for your ceaseless solidarity.

Gratitude, respect, and love to my family, particularly my mother and father, whose support is indefatigable and for which I am monumentally blessed. Hearty and heartfelt thanks to Lynn, Tracy, Tammy, Dean, and Tyler.

If you find anything of value in this book, credit should go to Joy Jensen, keenest reader and word slinger west of the Willamette—and east of it, too. Without her, this doesn't exist. Any shortcomings are mine entirely, and likely due to me not heeding her advice. Joy, my appreciation for you and your immeasurable help will be expressed in deeds. I love you madly.

But let's be honest. The dog deserves the most thanks. Digger, for literally sticking by my side while I write, *thank you.*

Contents

List of Figures

CHAPTER 1

Guilt as Keystone Emotion for Environmental Communication

Abstract This chapter introduces how guilt, as a keystone emotion for environmental communication, exerts a cascading influence on the composition and navigation of affective and rhetorical ecologies. It outlines the book's call for a richer understanding of environmental guilt and its collective dimensions, as well as the challenges presented in doing so. I propose an ecological approach to the study of guilt—in both metaphorical and material terms—because through improved engagement with environmental guilt we can be more effective in addressing our ecological crisis.

Keywords Environmental guilt · Rhetorical ecologies · Environmental communication · Affect · Emotion · Public feelings

Environmental guilt is often subtle. It might register as a flicker of pursed lips when tossing a to-go coffee cup in the garbage. It may manifest in a fleeting, faint press against your chest while you're paused at a stoplight, watching car exhaust billow up and vanish into the atmosphere. Its suasive forces may be at work when you scroll past headlines heralding the latest version of too-familiar stories: "threatened species continues to decline" or "impacts of climate change appearing faster than anticipated." Environmental guilt lodges in the slumped shoulders

T. Jensen, *Ecologies of Guilt in Environmental Rhetorics*,
Palgrave Studies in Media and Environmental Communication,
https://doi.org/10.1007/978-3-030-05651-3_1

of those who think *I should be doing more* and in nervous snickers over the "unseasonably warm weather" we've been having of late. We may not even notice any bodily signs of environmental guilt, even though its rhetorical forces are present, influencing our beliefs and behaviors in imperceptible, indirect ways.

Environmental guilt, after all, is suffused throughout our communication landscape. One need not experience acute distress and remorse over environmental harm to be affected by rhetorics that invoke it. In grocery store aisles, via news reporting, through activist campaigns, and in a multitude of other contexts, we are routinely prompted to evaluate environmental wrongdoing, assess who is at fault, and weigh our capacity for response. From recycling reminders to Facebook posts denying climate change to street-corner canvassers soliciting your save-the-polar-bear-from-extinction signature, contemporary environmental discourses bid audiences to consider their individual culpability in relation to a densely collective problem. These rhetorics generate a wide array of somatic reactions and cognitive maneuvers. Cavalier dismissal of ecological problems, defensive self-exoneration, and studied incuriosity—far from indicating a lack of engagement with environmental guilt—are often evidence of entanglement. Apathy in response to global warming frequently stems not from a deficit of awareness, as Kari Norgaard reveals in *Living in Denial: Climate Change, Emotions, and Everyday Life*, but from being *all-too-aware* of its significance. There are benefits, of course, to "avoiding the emotional and psychological entanglement and identity conflicts that may arise from knowing that one is doing 'the wrong thing,'" she writes.[1] The rhetorical reach of environmental guilt is extensive, stretching well beyond the conscious experience of guilty feelings.

That said, conscious feelings of guilt related to environmental issues are pervasive—and growing. National surveys suggest that "green guilt" is on the rise, with more than one-third of US respondents saying they feel badly that "they could and should be doing more to help preserve the environment."[2] "[If] I'm running late and I have to drive my car, I feel so bad about it," one student confesses, "cause I'm like, this is such a waste of gas … it's only me in the stupid little car to go five seconds down the road, I feel bad about it, but I have to get to class, so…"[3] The growing sense of environmental guilt is attributable in part to the fact you can also experience it for *in*action. "I feel guilt not just in environmental issues, like I'm worsening climate change or different things," a research participant admits, "but also I feel mostly guilty for *not* participating or being active."[4] Guilt is complex and contradictory: It can emerge not only from doing wrong but also from not doing right. It can arise for something you didn't do, perhaps out a sense of fear and confusion, as well as for something you didn't do, but *wanted* to. Guilt may be felt for something you *think* you did, even if you didn't. Conversely, you can *be* guilty of something and not *feel* guilty in the least. Such is guilt's plasticity, which enables its prevalence in the emotional ecology of our everyday lives.

Environmental guilt may feel different from other forms of guilt, since it is commonly one root thread in a complex braid of ecological uneasiness, apprehension, and feelings of helplessness. Research shows these feelings are widespread. In 2017, the American Psychological Association released a report on "Mental Health and Our Changing Climate," citing the growing impact of climate collapse and biodiversity decline on our psychological health. The report notes that clinical and scholarly research "stress[es] the possible detrimental impact of guilt, as people contemplate the impact of their own behavior on future generations."[5] It's not just adults who feel it. "Surveys conducted in different countries show that many young people are worried about climate change and rank the problem as the most important societal issue," notes psychologist Maria Ojala.[6] The "eco-anxiety" she observes in children and young adults coping with the knowledge of climate change, "is a worry mixed up with guilt."[7] Environmental guilt is an intergenerational phenomenon.

The desire to absolve oneself of these disquieting feelings is, unsurprisingly, equally prevalent. Guides for easing "eco-guilt" populate magazine columns with titles like "Offsetting Green Guilt" and "Eco-Guilt: Feeling Bad About Not Being Green."[8] GuiltyGreenie.com offers visitors a "non-judgmental and practical approach to change."[9] The book *Spit That Out! The Overly Informed Parent's Guide to Raising Children in the Age of Environmental Guilt* advises parents navigating an age replete with ethical dilemmas. "We are bombarded with new and contradictory research concerning environmental toxins, long-term product effects, and the far-reaching impact of every item we purchase and decision we make," it observes, adding, "[a]ll this information can feel like too much burden to handle."[10] Environmental guilt can motivate reparative behavior, but it can also act as a paralytic, trapping you in a loop wherein your desire for constructive action confronts options that all involve perpetuating some form of ecological destructiveness. Each encounter with environmental rhetoric, whether implicit or explicit, functions as a reminder to situate yourself within a crisis that is global in scope and unprecedented in scale. Everywhere, we are mired in sticky issues of complicity.

Keystone Emotion of Environmental Communication

Contemporary environmental rhetorics have greatly expanded awareness of the ecological dilemmas we're enmeshed in, all of which are interconnected—mass extinction, the breakdown of our climate, and pervasive pollution. Given how overwhelming the problems seem and how high the stakes are for addressing them, it's unsurprising that emotional turmoil and great uncertainty are spreading in breadth and gaining in depth. How we engage these forms of feeling has enormous impact on how we respond to ecological crisis. This book focuses on guilt as a pivotal factor for both the crisis and our emotional relations to it. The anemic response to ecological upheaval is due in large measure, I argue, to an inability to navigate novel forms of environmental guilt.

As a keystone emotion in contemporary environmental communication, guilt plays a critical role in the structure, circulation, and reception of environmental rhetorics because they are fundamentally mediated by ongoing ecological crisis. Even when not addressed explicitly, issues of culpability and responsibility resonate in the ambient of environmental rhetorics, influencing their interpretation and subsequent response. I adapt the concept of "keystone species" from ecological science to illuminate how guilt functions in relation to other emotions and behaviors, just as ecologists borrowed the term "keystone" from architecture to demonstrate how critical species relate to their ecosystems. In an arch, the keystone is placed at the peak center, exerting pressure on other pieces in a way that aligns the entire structure. In ecological science, a keystone species performs a similar function: it applies substantial influence across an entire ecosystem's structure and functioning, such as wolves do in Yellowstone National Park and elsewhere. As a keystone emotion for environmental communication, guilt exerts significant influence on how affective and rhetorical ecologies are organized and navigated. Who gets framed as culpable for environmental harm affects the flow of attention and resources; which actions are promoted as environmentally harmful will constellate how other actions are viewed; how feelings of guilt are interpreted regulates the available routes of redress. Guilt—as an emotional experience, cultural concept and catalyst, and rhetoric of allegation and atonement—impacts environmental communication at a constitutional level.

This book therefore attends to how guilt is provoked, perpetuated, and framed by everyday discourse about the environment. Combining affect theory with rhetorical analysis to examine a range of texts and media, I clarify some core challenges that guilt presents for communicating ecological issues of scale, particularly for rhetorics aimed at mobilizing others to take action. As a potent motivating force, guilt can prompt people to take reparative, ecologically restorative measures that alleviate the disquieting feelings and promote positive connections with others. For the same reason, guilt can also be marshaled to exploit others, as I show in analyzing how corporate industries wield guilt to both produce profit and constrain environmental activism.

These analyses reveal a clear need for building emotional literacies that more productively engage our complicity in global ecological harm. Emotional literacies are heuristics we use to parse feelings, detect their causes, and evaluate how they orient us toward some things and away from others.[11] We rely on these literacies to communicate our experiences and cultivate relationships with others, human and otherwise. The prevailing frameworks we use to make sense of guilt, however, have been calibrated to an *individualistic* notion of guilt. Environmental guilt, in contrast, is a form of *collective guilt*—feelings of distress and remorse for being part of a group that has committed unjust harm. This book tracks the rhetorical consequences and challenges of navigating environmental guilt when traditional frameworks render it largely illegible.

In taking a rhetorical approach to the study of emotion, this book also demonstrates that just as we are subject to guilt's persuasive energies, the concept of guilt is itself subject to rhetorical forces. What passes today as a completely natural notion of guilt has in fact been shaped over centuries by the vicissitudes of cultural, legal, religious, and philosophical rhetorics. Its current configuration—"as a property of individuals rather than collectives"—is in many ways anomalous within the wider scope of human history, since, philosopher Samantha Ashenden notes, for many societies, "guilt was both collective and objective."[12] In tracing its rhetorical history, we see how guilt has been molded by ideals of the Age of Enlightenment, which configured the individual subject as independent, self-willed, and guided primarily by the use of rational faculties.

Acknowledging that guilt has a past implies that it has a future. This book looks to a future where guilt and other emotions are shaped in tune with the ecologies that sustain us and others. Addressing ecological crisis involves recognizing the affective identifications and emotional frameworks that have facilitated that crisis—and then realigning them. How might guilt be reconfigured, for example, from an ecological view of the individual subject, who, far from being independent, is radically *interdependent*? What does responsibility look like if we swap the self-willed rational actor with the ecologically attuned being? What are the benefits of approaching emotions as *ecological*—not only metaphorically, but *constitutionally*, as material forces immanent within biophysical ecologies?

These questions call for reorienting our relationship to emotion as a category, to ecology beyond science, and to rhetoric as a discipline and everyday practice. While these ambitions obviously outstrip the bandwidth of these chapters, what follows is an attempt to advance us closer to such reoriented, rejuvenated modes of relating. By focusing analytical attention on the keystone emotion of guilt, my intention is to provide fertile foundation for those who also feel pulled to such matters of concern. This book is a call for scholars and activists to dedicate their creative and critical energy toward developing new emotional ecologies, ones that correspond to the new environmental realities unfolding before us, and that help write the unfinished story of ecological crisis.

Framing Environmental Guilt

In "Why It Matters How We Frame the Environment," George Lakoff recalls when he first learned that the earth's temperature would rise a full degree Celsius, perhaps two: "In seconds, my reaction was 'Omigod!' I had studied enough thermodynamics to know how huge an amount of heat that was [and] like other scientists, I believed at the time that if we all just got the scientific word out, the world leaders would see the threat and do the right thing."[13] His prediction, it turns out, was a bit optimistic.

Why have reports of climate breakdown and its catastrophic effects been met with such an anemic response? Why has the well-publicized plight of mass extinction stirred so little organized revolt against its causes? This is the central challenge—and frustration—for many studying environmental communication. Scientists and environmentalists have offered numerous explanations, the majority of which, notes Kari Norgaard, "emphasize either explicitly or implicitly the notion that information is the limiting factor in public nonresponse," a view encapsulated by the *information-deficit model.*[14] Correspondingly, the bulk of environmentalist rhetoric over the last half-century has been predicated upon *getting the facts out* and *raising awareness.* "This has been the Al Gore strategy," Lakoff explains, "if enough people ... are exposed to those facts, the world should change."[15] If people just knew what was happening, the reasoning goes, the only logical response would be to rise up and confront the issue. Once complicity is clear, people will act.

Without question, there has been some success with this strategy. Overall environmental awareness is likely the highest it has ever been in modern American history. The pressing question, however, is if this strategy is sufficient for adequately addressing the most urgent ecological issues. Substantial scientific documentation confirming the existence of global warming has circulated widely over the past several decades, as have reports that link the rapid escalation of global temperatures to human activities and the emission of carbon dioxide. During the time, however, fossil fuel use has increased and biodiversity has steadily decreased. Environmental awareness may be at its highest, but so is energy consumption, which hit a record peak in 2018.[16] Clearly, the prevailing approach of only providing information is insufficient. Nevertheless, prominent environmental campaigns continue to operate on the assumption that *knowledge* of environmental harm has a direct, causal relationship to effective *action* that addresses it. The relationship between factual knowledge and engagement, however, is not so tidy. Knowledge of ecological crisis can also cause *inaction*, as a risk-perception analysis discovered when it determined that "more informed respondents both feel less personally responsible for global warming, and also show less concern."[17] From a rhetorical perspective, the issue is not a matter of *more* or *less* environmental knowledge, but is instead a matter of *how* that knowledge is interpreted and acted upon.

The critical deficiency of an information-deficit model is its reductive understanding of the communication process. It flattens out essential elements of how we make sense of information, such as the contextual dimensions through which it is received, as well as how one's value systems and sense of personal identity color that knowledge. In reducing the complexities of communication and meaning-making, "public perceptions are seen as stable, coherent, and consistent and to exist within individuals, rather than being located within the inter-subjective contexts of institutions and discourse," notes political geographer

Harriet Bulkeley.[18] Over the past two decades, studies of environmental communication have consistently shown how inconsistent perceptions of ecological crisis are, and how deeply dependent they are on context and language choice. The phrase *climate change* evokes very different imagery and reactions than *global warming*.[19] Opinions on global warming are significantly influenced by local weather and fluctuate accordingly.[20] In some cases, the mere presence of dead indoor plants is enough to strengthen belief in global warming.[21] Information-deficit explanations of inaction often presume an independent, consistently rational individual, which is inadequate for illuminating how co-constituted actors within dynamic ecologies form their beliefs and pattern their behaviors. "One thing science communication scholars seem to agree on is the inadequacies of the deficit model," Alison Anderson points out in assessing key challenges in the field of environmental communication.[22] And yet, she adds, "scientists, government officials, and NGOs have made little use of the findings generated by scholars working in the science/environmental communication area."[23]

The concept of *framing* has been promoted as a remedy to an information-deficit approach and has gained widespread appeal in research on environmental communication over the past decade. Matthew Nisbet describes frames as "organizing devices" that every audience uses to make sense of information, situating it within "interpretative storylines" that simultaneously pare down complexity into more manageable forms while still endowing it with a sense of relevance.[24] Framing matters for public engagement on ecological issues, Nisbet argues, not only because they are "an unavoidable reality of the communication process," but also, through the carefully calibrated use of "researched metaphors, allusions, and examples," the right kind of framing can "break through the communication barriers of human nature, partisan identity, and media fragmentation [and] trigger a new way of thinking about the personal relevance of climate change."[25] Framing provides a richer understanding of the complexities of communication, and acknowledges what the information-deficit approach greatly undervalues: emotion and reason do not operate independently of one another. "Emotions are an inescapable part of normal thought," Lakoff notes.[26] "Indeed," he adds, "you cannot be rational without emotions."[27]

Framing highlights a simple fact of communication: facts do not speak for themselves. Information—particularly complex information about challenging topics, such as climate destabilization or mass extinction—is always interpreted through conceptual structures, which sort and schematize according to one's existing value systems. Environmental advocacy campaigns based on "getting the facts out" have largely been premised upon a mistaken understanding of human reasoning and communication. Synthesizing several decades of studies in neuroscience and cognitive psychology, Lakoff juxtaposes the assumptions of that faulty approach with framing, which is based on scientific research:

> Most of us were brought up with a commonplace view of how we think that derives from the Enlightenment. Over the past 30 years, the cognitive and brain sciences have shown that this view is false. The old view claimed that reason is conscious, unemotional, logical, abstract, universal, and imagined concepts and language as able to fit the world directly. All of that is false. Real reason is: mostly unconscious (98%); requires emotion; uses the "logic" of frames, metaphors, and narratives; is physical (in brain circuitry); and varies considerably, as frames vary. And since the brain is set up to run a body, ideas and language can't directly fit the world but rather must go through the body.[28]

Without first appreciating communication and reasoning as *embodied*, *emotional*, *contextual*, and *narrativized*, we will not be able to answer why rhetorics of ecological collapse have been met with such an anemic response. How we *feel* about environmental harm greatly influences how we *think* about it, and vice versa. Framing, as a concept and field of inquiry, is valuable in that it recognizes the reticulated relationships between rationality and emotion, logic and narrative, context and meaning.

Although framing offers a more sophisticated understanding of how environmental communication functions than an information-deficit perspective, research on the topic has nevertheless remained focused on the cognitive process, even as it declares the importance of emotion. As late as 2018, Nabi et al. write that "[t]he study of emotion in the context of framing is relatively recent, and as such, there is little theorizing."[29] It is perhaps more accurate to state that there has been little *explicit* theorizing of emotion within framing studies of environmental communication. On the whole, framing research implicitly individuates and instrumentalizes emotion. Guilt is seen as one tool among many in an inventory of emotions—most prominently, fear, hope, and anger—that can be deployed to move people to action (or inaction, depending on one's strategy). And because framing research largely delimits attention to "the construction of meaning in individual subjects," as Robert Cox contends, "people" often refers to an aggregation of individuals.[30] This approach is not unique to framing methodologies, but is rife across much of environmental communication research. As Norgaard notes, the "'conceptual challenges' surrounding global warming have been understood primarily in terms of limitations of individual psychology (i.e., mental models, confirmation bias) or of media framing."[31] The focus on individual limitations has itself become a limitation, leading Cox among others to call for "rethink[ing] the idea of climate communication at the scale of systems."[32]

How different environmental frames are cognitively processed on an individual level is valuable, yet largely elided from these conversations are the ways in which emotion is dynamically, rhetorically constructed and navigated on a *collective* level. By assuming the individual body as the natural site for the study of emotion, we tend to overlook the cultural, ecological, and historical contexts that exert enormous influence on our embodied engagement with the world. Moreover, when emotions are individuated in relation to each other, such that each represents neatly separated modes of experience, it becomes easier to see them as static, stable, ahistorical categories. We become, as a result, insensitive to how emotions are fundamentally intertwined, not only with each other, but also with the social structures within which we are enmeshed, as well as the ecological systems we are dependent upon.

While a turn toward framing has helped elevate the visibility and importance of emotions for both the study and practice of environmental communication, the research to date has implicitly cast them in rather restrictive and reductive ways. This book builds on the insights that framing research has helped catalyze, and similarly seeks to go beyond an information-deficit approach to provide more accurate and powerful explanations for why so little action has taken place, when so much is known about our ecological crisis. This book also seeks to go beyond framing and help incite a new orientation toward emotions, not only for the field of environmental communication and its methodological trends, but also and more importantly, for our everyday acts of environmental communication.

This reorientation necessarily entails recognition that reason and logic—far from being the opposite of emotion and embodied feeling—fundamentally rely on them. And this reorientation is already underway. While Lakoff cites the substantial amount of empirical evidence backing this claim, cognitive science is only but one branch of study that views emotion not as ancillary to inquiries of reason and behavior, but rather, as vital to them. Affect theory, a thoroughly interdisciplinary and wide-ranging area of study attending to issues of embodiment, sensation, power, and persuasion, has done substantial work over the past two decades to shift perspectives on how feeling factors into social and political phenomena. By combining affect theory with rhetorical theory, we can add ecological phenomena to that list. In doing so, we will be better equipped to answer why communication about climate breakdown and its ruinous effects have—so far—been met with a level of collective action that pales in proportion to the present problems.

From Environmental Awareness to Ecological Attunement

Simply increasing factual knowledge of ecological destruction is not the catalyst that many presume it to be. Many environmental rhetorics aim to place such knowledge into actionable contexts, and promote a sense of environmental awareness—that is, a routine recognition of and general alertness to the impacts of human activities on biophysical ecosystems. As one cultivates and expands this awareness, progression tends to yield a more enveloping *ecological attunement.*

The shift here is from seeing the environment as a setting or backdrop for human activity to seeing all things—including oneself—as immanently ecological, inescapably intertwined in webs upon webs of relationships. The etymological roots of environment allude to the sense of separateness that ecological attunement annuls: *environ*, from Old French *environer*, meaning the action or state of being surrounded or encircled. The move is from being encircled to being enmeshed. Attuning to that enmeshment involves ecological thinking, which Timothy Morton describes as "a practice and process of becoming fully aware of how human beings are connected with other beings—animal, vegetable, or mineral."[33] Ecological thought is more expansile than just thinking *about* ecology. It entails thinking *ecologically*, in which, as Lorraine Code writes, "[we] are repositioned as self-consciously part of nature, [at all times] reciprocally engaged, intra-active."[34] Like a virus that works to dissolve notions of sovereign bodies, ecological thinking exposes a radical integration with what one previously perceived as backdrop.

This seemingly slight shift in orientation—from being within an environment to being inseparably part of ecological processes—has profound ramifications for how one responds to ecological crises. After all, in almost perverse irony, it is often destruction or loss that catalyzes an awakening. Destabilization of biophysical ecosystems helps to illuminate the depths and details of our interconnectivity. Take, for example, the discovery that antidepressants are accruing in the brains of fish, the pharmaceuticals having been flushed out of our bodies and into local watersheds.[35] In a recent study, researchers found that mussels in the Puget Sound Bay tested positive for oxycodone, multiple varieties of antibiotics, antidepressants, and synthetic surfactants found in everyday cleaning products. They also discovered flame retardants, PCBs, pesticides, mercury, arsenic, an antidiabetic drug, and a chemotherapy agent.[36] It is one thing to learn about these discoveries, it is quite another to fully realize their implications. Attuning to such deep connections—between medications and mussels, for instance—alters the perception of everyday objects and processes, a transformation that only further accelerates sensitivity to the extent of ecological disruption.

This feedback loop is one reason why "[e]cological thinking reconfigures relationships all the way down: epistemological, ethical, scientific, political, rational, and other relationships between and among living beings and the inanimate parts of the world."[37] This reconfiguration highlights the impact that our choices—both individual and collective—have on ecosystems at multiple scales. Connections between actions and effects begin to bloom everywhere. The chicken in your Caesar salad suddenly transforms into one node of a much wider ecological web. You see not only the chicken, but also the crops that became its feed, the water and pesticides used to grow those crops, the massive amount of chicken manure produced by factory farms, the nitrogen from that manure contaminating watersheds, the algae blooms created by polluted run-off, and the fish kills triggered by the oxygen-less water that algae blooms leave behind; you might envision the fossil fuels used in transporting the crops to the mill, the feed to the poultry farm, the chickens to the slaughterhouse, and the meat to the grocery, where it is then packaged in plastic, an oil-based product; you might perceive all these connections and more—even if you'd prefer not to.

Whether passively developed or cultivated with great intention, ecological attunement entails amplified exposure to questions of culpability, which are the inevitable byproducts of recognizing connections across ecologies, economies, and cultures. *To what extent am I responsible for orangutans being pushed to extinction by eating this Girl Scout Cookie?*[38] *Am I connected to the decline of salmon just by doing a Google search?*[39] *Whose fault is it that microplastics are in my water, and in 94% of tap water sources in the United States?*[40] Such questions, of course, are almost never composed so cleanly in our daily negotiations. Most self-questioning of one's role in global environmental upheaval, I would wager, does not get formulated into conscious thought at all. Rather, of the many instances in which we are prompted to consider our connection to ecological crisis, the majority are registered primarily—and often only—at an affective level. "There you are," Morton writes, "turning the ignition of your car. And it creeps up on you. You are a member of a massively distributed thing."[41] These vague, often elusive, and generally disquieting feelings are the bodily registers of environmental transgression, and they are central to understanding how communication, identification, and persuasion occur in this moment of ecological catastrophe. We must acknowledge that new ways of thinking about the world are not without corresponding new ways of feeling in the world, particularly when that world is in turmoil.[42]

Ecological attunement, then, refers not only to ecological *thinking*, but also and importantly, to the affective and emotional energies that pattern the felt experience of ecological connection. Morton's *The Ecological Thought* and Code's *Ecological Thinking: The Politics of Epistemic Location* are two among many excellent books that articulate the awesome potential and critical necessity of recalibrating not just what we think *about* in our relations with the world, but *how* we think about them. Without attending to the emotional connections with ecologies that sustain us—materially and spiritually, but always literally—our understanding of ecological crisis will be impoverished. These are the forces that inform our values, influence our beliefs and behaviors, and that subsequently move us, just as the etymology of *emotion* intimates, with its Latin root of *ēmovēre* meaning to stir or disturb into motion. It is not enough for scholars, students, activists, and advocates to only affirm the importance of affect and emotion for ecological relations; we must also explore and experiment with these vital, formative connections, while encouraging others to do so as well. The creative and critical work that results will help us confront the pressing problems of ecocide, including why communicative efforts have so far elicited comparatively little collective, effective action. This work can also equip us rhetorically for navigating the novel forms of feeling and thinking that are being animated by sweeping planetary change. Above all, approaching affect and emotion from an ecological perspective can yield transformative, reparative relationships with a world under siege. In moving from environmental awareness to a more immersive ecological attunement, our sense of rhetoric must shift too—a metamorphosis that's already begun.

Ecologies—Rhetorical, Affective, and Material

Over the past fifteen years or so, a body of scholarship has matured around various articulations of rhetoric and ecology: ***rhetorical ecologies*** (Edbauer, 2005), ***ecological rhetoric*** (Ells, 2008), ***rhetoric's ecologies*** (Wells et al., 2017), ***rhetorical ecologist*** (Druschke, 2019).[43] "[R]ichly ecological perspectives have been forwarded to explore connections between interdependency and addressivity," Wells et al. note in the introduction to their compelling collection, *Tracing Rhetoric and Material Life: Ecological Rhetorics*. As a consequence of these perspectives, they add, "rhetoric as traditionally conceived has become more networked, material, hybrid, immersive and consequently dynamic."[44] An ecological approach to rhetoric, in other words, emphasizes the embodied and the embedded, is sensitive to connection and circulation, and directs attention to how rhetorics adapt and evolve across contexts. I say that this thread of inquiry has "matured"—rather than, say, emerged—because rhetoricians and compositionists have long been interested in ecology, whether as a model for the composition process, as Marilyn Cooper proposed in 1986, or for its rhetorical attributes, as R.L. Scott suggested in 1973.[45] True to form, Kenneth Burke is eerily prescient in this regard, writing in 1937, "there is one little fellow named Ecology, and *in time we shall pay him more attention*."[46] Rhetoric studies *is* paying more attention to ecology (and pronouns) and the work being produced is increasingly sophisticated and enlivening. This trend is poised to continue, too, as more scholars and students attune to the ethical imperatives of our ecological present.

Burke's augury of ecology was borne from observations of the Dust Bowl and its perspective-shifting aftermath. "[The] laws of ecology have begun avenging themselves," he writes, "against restricted human concepts of profit by countering deforestation and deep plowing with floods, droughts, dust storms, and aggravated soil erosion."[47] These geophysical impacts led Burke to begin "using the term 'ecological balance' *literally*," in addition to his figurative treatments.[48] Like Burke, our increased attention to ecology in the field of rhetoric stems from multiple desires, and includes an appreciation for the methodological and stylistic maneuvers that an ecological perspective animates. The unsettling recognition of current crises and their corresponding exigencies, however, plays no small role in this growing body of work. These mutualistic impetuses—of metaphorical and material ecologies—have informed scholarly methods, resulting in a rich and reticulate understanding of what an ecological approach to rhetoric might entail.

Attempting to articulate a precise and permanent notion of what an ecological approach to rhetoric "is," however, misses the point. As Wells et al. rightly note, "the trope of ecology is neither singular nor static—rather it is itself ecological or tropic, 'continuously reconfigured through repetition and difference, expressing rhetoric's creative potential'."[49] If there is a unifying notion to ecological rhetorics, it is a resistance to a single unifying notion. Ecological approaches accentuate the adaptivity, dynamism, and contingency that constitute systems of action and exchange. They are characterized by an abiding interest in qualities of interrelation, circulation, and the flux and flows of affective energy; they investigate enmeshment, open networks, and systems across scales; they attune to embodiment and porosity, immersivity and addressivity, entanglement and evolution. And from within this embrace of complexity, ecological approaches attend to patterns and relationships of rhetorical energies.

When I turn this approach on itself, I see pronounced concern in certain clusters of inquiry—affect, materiality, adaptivity, and systems—and yet also see how these are ultimately not separate categories, but deeply intertwined and co-constitutive. It is precisely because of, not in spite of, the energetic exchanges between these permeable categories that I find in rhetorical ecologies a fruitful orientation for tracing guilt's suasive forces. The ecological approach that I adopt in the chapters that follow, then, attends to *systems* instead of singularities, positions *contingency* over certainty, views complex *entanglements* as the presumed norm, and intuits *affect* as an energy inherent to material ecologies.

Jenny Rice outlines such an approach in contrasting rhetorical ecologies" with the rhetorical situation, as conceived by Lloyd Bitzer. "Rather than imagining the rhetorical situation in a relatively closed system," Rice writes, "it is also possible to situate [it] within a wider context of *affective ecologies* comprised of material experiences and public feelings."[50] In reframing the rhetorical situation from a collection of discrete elements (e.g., speaker, ethos, audience, etc.) to a dynamic set of interrelations always teeming with potential, Rice's argument "inaugurated an abiding interest in the 'rhetorical ecologies' of public rhetoric," note Wells et al.[51]

Similarly, Rice's use of "affective ecologies" demonstrated that affect theory also finds a natural partner in ecological orientations, along with rhetorical theory. Though the scholarly methods, data, and outcomes of ecological science are obviously quite different, there is an uncanny overlap in the overarching issues on which their attention is placed. The Ecological Society of America locates the discipline's fundamental interest in the "physiological responses of individuals, structure and dynamics of populations, interactions among species, organization of biological communities, and processing of energy and matter in ecosystems."[52] By comparison, in their introduction to *The Affect Theory Reader*, Gregory Seigworth and Melissa Gregg write,

> affect is integral to a body's perpetual *becoming* (always becoming otherwise, however subtly, than what it already is), pulled beyond its seeming surface-boundaries by way of its relation to, indeed its composition through, the forces of encounter. With affect, a body is as much outside itself—webbed in its relations—until ultimately such firm distinctions cease to matter.[53]

In both the study of ecology and affect, focus is on energy, exchange, and circulation. The same can be said for ecological approaches to rhetoric. Highlighting the facility with which elective affinities between affect, rhetoric, and ecology can be found is not to suggest there aren't distinctions between them. Rather, it is to acknowledge the potent explanatory capacity their connections create.

Affect resists easy definition, just as ecology and rhetoric do. I use the term affect to indicate the energies that circulate between bodies in relation, naming those invisible, yet palpable forces that modify bodies—increasing or diminishing, helping or hindering their capacities. These relational forces are always more-than-human, as affect is inherent to material ecologies. The energies that animate us are not exclusive to human bodies. As Jane Bennett suggests, "*All* forces and flows (materialities) are or can become lively, affective, signaling. And so an affective, speaking human body is not *radically* different from the affective, signaling nonhumans with which it coexists, hosts, enjoys, consumes, produces, and competes."[54] We are thoroughly, irreducibly interdependent. We must eat food and drink water and breathe air to remain alive. In other words, what connects us is our composition.

An ecological approach to rhetoric—one that furls together questions of affect, biophysical materialities, interdependency, and rhetoric—invites reconsideration of some of the most established concepts and theoretical assumptions in the field, to the very ontological foundations of rhetoric itself. In *Inessential Solidarity*, Diane Davis contends it is past time "for rhetorical scholars to push beyond the (merely) epistemological concerns that have for so long circumscribed our theories of persuasion" and reconsider the boundaries of "who or what might be engaged" in rhetorical relations.[55] Bridie McGreavy and Nathan Stormer answer that call, asserting that "[r]hetoric's ontology, approached ecologically, considers *qualities of relations* between entities, not just among humans, that enable different modes of rhetoric to emerge, flourish, and dissipate."[56] Caroline Gottschalk Druschke advocates for a more radical renewal, calling for a rhetoric that is "a connective verb composed of physical, palpable, symbolic, affective, and chemical relations….rhetoric *as ecology*. And ecology *as rhetoric*."[57] As Burke predicted, rhetoricians are paying more attention to ecology, and prompting paradigm-shifting questions as a result. I argue that the study of emotion—from an ecological, rhetorical stance—is essential to extending rhetoric beyond the social, epistemological, and human-centric realms to which it has traditionally been bound. There is an urgency to such work, too, as we navigate our way through a rapidly changing planet.

Emotional Ecologies

For many of us, a dim sense of feeling bad about environmental issues pulsates in the background most days. Like the low, constant hum of a refrigerator, it becomes part of an everyday atmosphere, perceptible but not recognized, unwelcome yet familiar. Precisely because its persistence renders it inconspicuous, it accompanies our routine actions and is managed with well-practiced coping mechanisms. We become accustomed. Only periodically are we pierced by moments of clearly perceived and fiercely felt environmental guilt. The phenomenological purview of environmental guilt, in other words, is as expansive as it is equivocal.

Conceptually, phenomenologically, and rhetorically, guilt is protean. When we look to pin down guilt as a concept, we find it entangled within legal, moral, political, spiritual, and phenomenological contexts, operating within individuals and among collectives, and effecting a range of actions, behaviors, and motives. As noted at the outset, guilt can emerge from doing wrong, but just as easily from not doing right. One can be found legally guilty but not feel guilty. The inverse is true as well. Guilt can range from intensely personal to diffusely collective, individual to intergenerational. These intricacies demand an approach to its study that accounts for such dynamism. An ecological orientation befits such complexity. The same set of methodological suppositions that has spurred powerful insights for rhetorical theory and environmental communication is equally apt for the study of emotion.

As a keystone emotion for environmental communication, guilt exerts a cascading influence on how related affective and rhetorical ecologies are organized and navigated. By analyzing *ecologies of guilt*, attention is directed to the myriad interrelationships among emotions, as well as to the contexts that animate and inform those relationships. The broader category implied by "ecologies of guilt"—namely, emotional ecologies—connotes this co-constitutive quality of emotions. Just as there is a danger to conceiving of ethos, pathos, logos as discrete elements of a lived situation, or sender, receiver, message as separate components that can be cleanly compartmentalized, so too is there a risk in believing the affective frequencies of shame can be neatly severed from those of guilt, for example. To presume that emotions are intricately connected and dynamically evolving is a means of respecting our subject matter, acknowledging our methodological limitations, and appreciating emotions as powerful forces whose full functioning will always exceed complete understanding.

I am, of course, certainly not the first to analyze emotions from an ecological stance. The Stoics, for example, adopted natural science schemas to discern emotions, inventorying them in terms of genus and species.[58] The metaphorical ease and intuitiveness of ecology is part of what makes this approach attractive, as it has potential for generating emotional literacies that can simultaneously express nuance and feel accessible in affectively vibrant ways. In *Carbon Footprints as Cultural-Ecological Metaphors*, Anita Girvan goes further, arguing that ecological metaphors bear traces of nonhuman forces and "communicate the agency of larger-than-human actors who come to challenge a given human way of constructing the world."[59] The provocative links between metaphor, affect, and materiality indicate to me that critiques of ecology being used "merely" as a metaphor greatly undersell both the power of metaphor as rhetorical construct and the catalytic capacity of ecology as metaphor.

In addition to the methodological and metaphorical advantages, I utilize emotional ecologies with a larger goal in mind, one that aims to shift how we conceive of emotion at a foundational level. Both explicitly and implicitly, emotions have long been framed as social phenomena. This book argues for treating emotions as ecological—not only metaphorically, but also constitutionally—as forces immanent within biophysical ecologies. This is not to deny nor diminish the sociocultural shaping of emotions. Indeed, this book largely focuses on how guilt has been molded by cultural forms and the consequences this has for the health of our planet. Rather, to claim that emotions are ecological is to recognize that our emotional health is inextricably tied to the health of our biospheres. *This is not a metaphor*. A substantial body of evidence exists that links environmental destruction with physiological health effects, even when excluding direct toxification. There is also ample evidence showing a positive correlation between time spent in greenspaces and biodiverse natural areas with improved attention, decreased stress levels, improved digestion, memory recall, and overall mood.[60] Though scientific studies and empirical data are helpful, they are not necessary for realizing that emotions are ecological. If you love a landscape and then that land and its inhabitants are violently abused, you will feel strong reactions. To accept that emotions are ecological is not to learn something new, but to remember something ancient.

To root emotions in material ecologies might invite reductive, binary frames that pit biology against culture, however, so a point of clarification is warranted. As Kay Milton succinctly notes, "[the] problem with [biological] essentialist or positivist approaches to emotion is that they emphasize the feelings but tend to ignore the [cultural] meanings, whereas constructionist approaches focus on the meanings at the expense of the feelings."[61] Like Milton, I argue that these extremes can be avoided if we recognize that emotions "are induced when an organism interacts with objects in their environment."[62] For many, that environment is saturated with social relations with other humans, but it is critical to remember "our environment is not essentially or primarily social; it is simply an environment."[63] To see emotions as ecological is to undermine numerous deep-seated and pernicious binaries, and invite reframing of the relationships between rationality and emotion, nature and culture, human and nonhuman.

An ecological approach bypasses the false either/or choices set up by those binaries and aids us in seeing emotion as rhetorical energies within material ecologies. Emotions are rhetorical in the sense that they influence how we relate, what we believe, and the ways we act. But they are also rhetorical by virtue of being the *subject* of rhetorical forces themselves. Cultural conceptions of emotion shape normative modes of feeling, privilege the experience of certain emotions over others, suggest and sanction the expression of emotions, and configure literacies for identifying, observing, and navigating those experiences. The boundaries of emotion are rhetorically crafted through the terms we use to communicate them, and the depth of vocabularies for emotion can greatly shape the experience of and consequences resulting from certain emotions.

The Chinese language, for example, has over one hundred words for shame. In contrast, the English language is comparatively impoverished in articulating the valences of shame—just one example of how vocabularies often reflect cultural hierarchies of emotion, organizing mores and rituals around certain emotions while minimizing others. "In Western cultures," Herant Katchadourian argues, "shame has occupied a position secondary to guilt as a moral emotion, whereas the opposite is true in Asian culture."[64] Thus, while an ecological orientation reminds us that emotions are not reducible to culture, we should also never underestimate the role culture plays in shaping emotional experiences—particularly in regard to expressing ecological connection. "Cultures all over the world have concepts in their language that relate psychological states to states of the environment," Glenn Albrecht writes.[65] Despite the well-established links between environmental and mental/physical health, Albrecht notes that "we have very few concepts in English that address environmentally induced mental distress and physical illness."[66] We need revitalized emotional literacies and vocabularies adapted to the ecological realities of our age. An ecological orientation to emotion can help accomplish this, providing a framework for analyzing the rhetorical and emotional mechanisms that have contributed to the precarious state of ecosystems worldwide, as well as for attuning to new ways of feeling that don't rely on harmful binaries.

The insights generated by affect theorists over the past several decades offer a strong foundation for analyzing emotional ecologies. Much of this work has sought to move the site of emotions away from its conceptual confines in the individual body and to rethink the cultural and political construction of feelings. Sara Ahmed, Lauren Berlant, and Ann Cvetkovich offer powerful analyses of how dominant discourse tends to psychologize emotion, inscribing it as a uniquely individual possession. As Ahmed notes, "the everyday language of emotion is based on the presumption of interiority."[67] In *The Secret History of Emotion*, Daniel Gross points to the roots of this individualist framework, noting that in the wake of a seventeenth-century turn to empiricism and the rationalist influence of Rene Descartes, "emotions that were once treated by everybody as externalized forms of currency and worldly investments [got] sucked, as it were, into the brain."[68]

A number of consequences flow from theorizing emotion at the level of the individual body, whether psychobiologically, where explanations of emotions are regularly reduced to chemical reactions, or psychologically, where emotions are often framed as internally forged, sublimated, or repressed. In both, emotion is figured as coming from within and then expressed outwardly, emanating from the body. Consequently, "solving problems [of emotion and desire] is the individual's responsibility," as Berlant notes. When emotion is framed as a psychological state of mind and the individual becomes the primary locus for emotional agency and accountability, attention to the social dimensions of emotion wanes. As a result, we become less aware of the ways in which our cultural environment shapes possible interpretations of emotions, how our mediascapes influence what becomes an object for emotional attachment, and how social norms are established through and buttressed by affective forces. Critiques of individualized models of emotion "remind us of the importance of understanding emotions not as psychological dispositions, but as investments in social norms," argues Ahmed.[69] Because ecocide could not occur without mass emotional disorientation, it is imperative that we identify how we become emotionally invested in norms that allow ecocide to continue. We must also, however, be more explicit about the fact our emotions are directly tied to the ecologies that sustain us and others. These critiques can help fashion new emotional literacies that begin from the well-established truth that our health is intimately interconnected with the health of the land and sea. It is time for an ecological turn in affect theory.

Guilt in Environmental Communication Scholarship

"The first decade of the twenty-first century may be considered an inflection point for scholarship in environment and communication," propose Anders Hansen and Robert Cox in opening *The Routledge Handbook of Environment and Communication*. As they elaborate:

> In asking whether such scholarship was a "crisis discipline," the inaugural issue of *Environmental Communication: A Journal of Nature and Culture* (2007) posed explicitly, for the first time, the question of why, and in what ways, studies investigating the nexus between the environment and human communication constituted an academic "field" as such.[70]

Of the two embedded questions here, the latter—*is environmental communication a recognized academic field?*—was settled swiftly in the inaugural issue, with plenty of evidence to answer affirmatively. The first question—*is environmental communication a* crisis discipline?—registered more debate. In fact, the first issue of the field's principal journal is dedicated almost exclusively to questions of ethical duty and advocacy in environmental communication scholarship.

The impetus for the first question comes from the same source of exigency that spurred the field into being. Robert Cox describes it as "a moment of conjunctural crisis, defined in not insignificant ways by human-caused threats to both biological systems and human communities, and also by the continuing failure of societal systems to sufficiently engage these pressures."[71] Using conservation biology as a comparison, Cox recommends being explicit and transparent about *why* we're doing this work: ecological crisis is caused by humans; humans are responsible for averting further harm and repairing the harm already done; we can still doing something about it; and as professionals, we can make positive contributions through our research.[72] Although some scholars offer cautionary gestures regarding methodology and reminders about language's limits, no rebuttals are made to the underlying logic: the field of environmental communication is a collective response born from an abiding concern for the human-induced ecological upheaval and the widespread failure to sufficiently respond, despite being collectively culpable for the harm and responsible for addressing it.

These axioms of the field are themselves rhetorics of environmental guilt. They are expressive recognitions of having done wrong, of having failed to rectify that wrong, and a sense of obligation to repair the relationships wounded by the offense. From the beginning, scholarship in environmental communication has emanated from—and often explicitly cites—a sense of responsibility to improving a situation in which scholars see themselves implicated. Moreover, a substantial portion of that scholarship examines the efforts of others (activists, NGOs, etc.) who identify as part of a collective committing ecological harm and who therefore feel compelled to redress those wrongs and obviate future harm.

Given these foundational motivations, it is perhaps with some surprise that the topic of guilt has received scant attention. No monographs, nor edited collections have taken on environmental guilt as their primary topic. Collective guilt—awareness that a group one belongs to has committed unjust harm—and its relationship to ecological destruction has received even less attention. One must look to other fields, such as environmental philosophy and environmental psychology, to discover work on collective guilt, where it nevertheless remains a topic of infrequent attention. Some degree of gap exists, then, between (1) the field's foundational acknowledgements of collective culpability and sense of responsibility, and (2) the extent to which those animating forces have been analyzed and theorized within existing scholarship. This book aims to prompt more conversation about those forces, so that we can better understand how they inform our beliefs and behaviors.

Although the study of guilt within environmental communication may be spare, it is not entirely absent. In "Between Guilt and Obligation: Debating the Responsibility for Climate Change and Climate Politics in the Media," Post et al. perform a robust quantitative content analysis of international news coverage regarding causal responsibility (blame) and treatment responsibility (obligation) for climate change between developed and emerging countries. Statistically significant results indicate that between 2004 and 2014, "media in developed countries … shifted responsibility for tackling climate change away from their in-group and to their out-group," which suggests that "transnational conflicts over climate change might not decrease but aggravate."[73] In another study, psychologists Swim and Bloodhart analyze data of study-participants who rated 13 different emotions on a 5-point scale in response to a range of environmental issues, particularly "messages portraying polar bears harmed by climate change."[74] Contrary to arguments that say such images will upset audiences and deter them from engagement, their results suggest that such portrayals "motivated both environmentalists and non-environmentalists," and from this they extrapolate that "the activation of guilt can potentially be effective at motivating climate change activism."[75] This conclusion aligns with several other psychology studies that find an increase in "eco-guilt" can increase the likelihood of "pro-environmental behavior," a topic I discuss further in the following chapter.[76]

The conclusions drawn by Swim and Bloodhart are complicated by the observations of Dannenberg et al. in their article, "The Moral Appeal of Environmental Discourses: The Implication of Ethical Rhetorics."[77] Bringing together methodologies from sociolinguistics and rhetorical studies to analyze interviews conducted with students in the "Earth Sustainability" undergraduate program at Virginia Tech, their conclusions indicate that moral and ethical discourses on sustainability can actually be counterproductive. Students are induced to experience guilt about engaging in environmentally harmful behavior (e.g., driving to school), they note, but they are also constrained by a lack of alternative means (public transit) and other obligations they feel compelled to meet (arriving to class on time, pleasing their parents, etc.). "The moral appeal appears to create a rhetorical double bind," the authors observe, and the result "is both inaction and guilt: the individual does not feel that she has the power to change either social expectations or her own desires."[78]

These findings offer a compelling reminder that the rhetorical dynamics of environmental guilt are complex and contingent. What constitutes "pro-environmental behavior" is never a static, nor politically neutral category. The lived reality of environmental guilt involves cognitive dissonance and the emotional entanglements it generates takes a toll on those who engage with environmental issues and reflect on their actions, with short- and long-term effects on behavior. It demands careful study from a range of disciplinary perspectives, but particularly those with rhetorical sensitivity attuned to its persuasive complexities. While each of these three studies offer valuable insights, they also suggest how comparatively thin the archive on environmental guilt is, given how pivotal it can be as a motivating force, as illustrated by the students' narration of everyday concerns and by the field's founding impetus.

Although a rhetorical approach to affective and emotional ecologies is not yet prevalent within environmental communication studies, there is a detectable eagerness for it. In their editorial essay for a special issue of *Environmental Communication*, Christensen et al. offer "a final thought to point to future research horizons" and suggest that the field would profit from increased attention to affect.[79] In light of "the degree to which popular narratives can inform values and citizen or policy action," the issue's editors write, "the study of affect could be considered as one of the central aspects in research agendas that aim to understand the relation between popular communication and societal change."[80] This call echoes the principal finding of Alex Lockwood's survey of affect and emotion research in environmental communication. Inventorying a broad body of scholarship on emotion and ecological issues, he finds that most are "engaged with individual values," "threatened identities," or "visual information's ability to arouse emotions."[81] "[The] actual ways in which affect works in and through social systems," however, "remains under-theorized and under-employed."[82] "[W]ithout intersectional knowledge of the theories of affect offered in other disciplines," Lockwood argues, "the study of environmental communication will be unable to fully get to grips with the pivotal role of emotions within environmental communication."[83]

For moving the field forward, Lockwood recommends turning to many of the same affect theorists cited in the previous section, who are loosely gathered under the banner of *public feelings*, a strain of scholarship focused on "understanding how everyday experience is a manifestation of social life," as Ann Cvetkovich puts it in *Depression: A Public Feeling*.[84] Taking everyday affective patterns as "both subject and method" of study, analyses of public feelings aim "to find new ways of articulating the relation between the macro and micro" in structures of power, investigating "*how capitalism feels*," for example.[85] "[We] do not always register that low-level fear and fatigue of living under capitalism," Lockwood notes, adding, "This is the affective register of climate change, also, and environmental threat."[86]

This book is my effort to sensitize us to the everyday vibrations of environmental guilt, a feeling of significant political and ecological consequence. It contributes to the larger project of constellating public feelings, and aspires to offer new ways of approaching guilt as others have offered fresh views on compassion (Berlant, 2004), depression (Cvetkovich, 2012), fear (Ahmed, 2004), pride (Gould, 2009), optimism (Berlant, 2011), and happiness (Ahmed, 2010), while humbly recognizing that such august company sets a high bar.[87]

This book is also a call for the sustained study of environmental guilt and its rhetorical influence on our engagement with ecological crisis.

The chapters that follow serve as a point of departure for this broader inquiry. In "Guilt's Plasticity," I outline key challenges for assigning, experiencing, and representing guilt. Looking at its rhetorical history, I argue, yields insight into why we are so ill-equipped to navigate its collective forms. In the third chapter, "Eco-friendly Scapegoats," I expose some of the consequences of our impoverished frameworks for grappling with collective guilt. This chapter outlines how guilt is weaponized through rhetorics of consumption to shape norms of what constitutes effective action against environmental harm. In chapters four and five, I trace guilt's entanglements with shame and grief, respectively. In "Guilty of Shame in the *Anthropocene*," I argue that the key term "Anthropocene" attributes global environmental harm to humans as a *species*, orienting us toward shame and away from taking accountability for the specific actions causing the harm. In "Guilty Grief and Ecological Mourning," I argue that feelings of environmental guilt threaten grief's transformation through the rhetorical work of mourning, indicating the need to rethink these emotions in tandem, as connected elements in an emotional ecology.

The book's epilogue, "The Future of Environmental Guilt," looks briefly at what tools and theoretical reorientations are needed to fashion new emotional literacies that better align with the affective realities we face in an age of ecological crisis. Rather than end on a note of hope, as so many environmental arguments calling for change are wont to do, I contend that hope need not be the only motivating emotion we use to accomplish the work of ecological attunement. If we render guilt edifying, and open ourselves to it as a teacher and source for revealing and refining our values, we can expand our emotional capacities and engage our ecological work with more energy, clarity, and efficacy.

Notes

1. Kari Marie Norgaard, *Living in Denial: Climate Change, Emotions, and Everyday Life* (Cambridge: MIT Press, 2011): 72.
2. "'Green Guilt' on the Rise," *Call2Recycle*, https://www.call2recycle.org/green-guilt-on-the-rise.
3. Qtd. in Clare Dannenberg, Bernice L. Hausman, Heidi Y. Lawrence, and Katrina M. Powell, "The Moral Appeal of Environmental Discourses: The Implication of Ethical Rhetorics," *Environmental Communication* 6, no. 2 (2012): 224.

4. Qtd. in Norgaard, *Living in Denial*, 195. Emphasis added.
5. Susan Clayton, Christie Manning, Kirra Krygsman, and Meighen Speiser, *Mental Health and Our Changing Climate: Impacts, Implications, and Guidance* (Washington, DC: American Psychological Association, 2017): 27.
6. Maria Ojala, "Eco-anxiety," *RSA Journal*, no. 4 (2019): 11.
7. Ibid., 12.
8. Matthew J. Kotchen, "Offsetting Green Guilt," *Stanford Social Innovation Review* 7, no. 2 (2009); Brooke Kosofsky Glassberg, "Getting Over Eco-Guilt," *The Oprah Magazine*, May (2008).
9. Home page, *The Guilty Greenie*, www.guiltygreenie.com.
10. Paige Wolf, *Spit That Out!: The Overly Informed Parent's Guide to Raising Healthy Kids in the Age of Environmental Guilt* (Gabriola Island, BC: New Society Publishers, 2016): 2.
11. For more on the term's origins, see Alison Waterhouse, *Emotional Literacy: Supporting Emotional Health and Wellbeing in School* (New York: Routledge, 2019); Amy E. Winans, "Cultivating Critical Emotional Literacy: Cognitive and Contemplative Approaches to Engaging Difference," *College English* 75, no. 2 (2012).
12. Samantha Ashenden, "The Persistence of Collective Guilt," *Economy and Society* 43, no. 1 (2014): 55, 56.
13. George Lakoff, "Why It Matters How We Frame the Environment," *Environmental Communication* 4, no. 1 (2010): 78.
14. Norgaard, *Living in Denial*, 1.
15. Ibid., 79.
16. U.S. Energy Information Administration, "In 2018, the United States Consumed More Energy Than Ever Before," April 16, 2019, https://www.eia.gov/todayinenergy/detail.php?id=39092/.
17. Paul Kellstedt, Sammy Zahran, and Arnold Vedlitz, "Personal Efficacy, the Information Environment, and Attitudes Toward Global Warming and Climate Change in the United States," *Risk Analysis* 28, no. 1 (2008): 113.
18. Harriet Buckeley, "Common Knowledge? Public Understanding of Climate Change in Newcastle, Australia," *Public Understanding of Science* 9 (2000): 316.
19. Anthony Leiserowitz, Geoff Feinberg, Seth Rosenthal, Nicholas Smith, Ashley Anderson, Connie Roser-Renouf, and Edward Maibach, *What's in a Name? Global Warming vs. Climate Change*. Yale Project on Climate Change Communication (New Haven: Yale University, 2014): 1–32.
20. Simon Donner and Jeremy McDaniels, "The Influence of National Temperature Fluctuations on Opinions About Climate Change in the U.S. Since 1990," *Climatic Change* 118, nos. 3–4 (2013): 537–550.
21. Nicolas Guéguen, "Dead Indoor Plants Strengthen Belief in Global Warming," *Journal of Environmental Psychology* 32, no. 2 (2012): 173–177.

22. Alison Anderson, "Reflections on Environmental Communication and the Challenges of a New Research Agenda," *Environmental Communication* 9, no. 3 (2015): 380.
23. Ibid.
24. Matthew Nisbet, "Communicating Climate Change: Why Frames Matter for Public Engagement," *Environment: Science and Policy for Sustainable Development* 51, no. 2 (2009): 15, 18.
25. Ibid., 15.
26. Lakoff, "Why It Matters," 72.
27. Ibid.
28. Ibid.
29. Robin L. Nabi, Abel Gustafson, and Risa Jensen, "Framing Climate Change: Exploring the Role of Emotion in Generating Advocacy Behavior," *Science Communication* 40, no. 4 (2018): 449.
30. Robert J. Cox, "Beyond Frames: Recovering the Strategic in Climate Communication," *Environmental Communication* 4, no. 1 (2010): 130.
31. Norgaard, *Living in Denial*, 65.
32. Cox, "Beyond Frames," 130.
33. Timothy Morton, *The Ecological Thought* (Cambridge: Harvard University Press, 2012): 7.
34. Lorraine Code, *Ecological Thinking: The Politics of Epistemic Location* (Oxford: Oxford University Press, 2006): 33.
35. Prapha Arnnok, Randolph R. Singh, Rodjana Burakham, Alicia Pérez-Fuentetaja, and Diana S. Aga, "Selective Uptake and Bioaccumulation of Antidepressants in Fish From Effluent-Impacted Niagara River," *Environmental Science & Technology* 51, no. 18 (2017).
36. Jennifer Lanksbury, "Stormwater Action Monitoring 2015/16 Mussel Monitoring Survey: Final Report," WDFW Reports, http://wdfw.wa.gov/publications/01925/.
37. Code, *Ecological Thinking*, 480.
38. Most Girl Scout Cookies are loaded with palm oil, the production of which is a principal driver of deforestation in Southeast Asia. Slash-burning millions of acres has devastating consequences for orangutan populations, which have sharply declined in the face of habitat theft and development.
39. According to Google, each search translates to roughly .2 g of carbon dioxide. By comparison, boiling a full kettle of water on the stove takes roughly 7 g—or 35 Google searches. If you live in the Pacific Northwest, as I do, chances are high that your electricity is generated by hydropower dams that block the passage of wild salmon returning to spawn. Google recently built another data center in Oregon, drawing power from the Columbia River. In 2000, their site averaged 18 million searches per day. In 2018, they averaged 228 million searches per hour.

40. A study of plastic fibers in tap water conducted by Orb Media found the United States had the highest concentrations: 94% compared to the global average of 83%. See "Invisibles: The Plastic Inside Us," by Dan Morrison and Chris Tyree.
41. Timothy Morton, *Dark Ecology* (New York: Columbia University Press, 2016): 8.
42. This is not to suggest that ecological attunement itself is novel, as numerous cultures have illustrated it is neither new nor the exception. What is new, however, is the severity of planetary change.
43. Jenny Edbauer (Rice), "Unframing Models of Public Distribution: From Rhetorical Situation to Rhetorical Ecologies," *Rhetoric Society Quarterly* 35, no. 4 (2005); Kevin Ells, "Ecological Rhetoric Through Vicarious Narrative: The Enduring Significance of Garrett Hardin's *The Tragedy of the Commons*," *Environmental Communication* 2, no. 3 (2008); Justine Wells, Bridie McGreavy, Samantha Senda-Cook, and George F. McHendry, Jr., "Introduction: Rhetoric's Ecologies," *Tracing Rhetoric and Material Life: Ecological Approaches*, eds. Bridie McGreavy, Justine Wells, George F. McHendry, Jr., and Samantha Senda-Cook (New York: Palgrave, 2018); Caroline Gottschalk Druschke, "A Trophic Future for Rhetorical Ecologies," *Enculturation: A Journal of Writing, Rhetoric, and Culture* 28 (2019).
44. Wells et al., *Tracing Rhetoric*, 3.
45. Marilyn M. Cooper, "The Ecology of Writing," *College English* 28, no. 4 (1986); Robert L. Scott, "On *Not* Defining 'Rhetoric'," *Philosophy and Rhetoric* 6, no. 2 (1973).
46. Kenneth Burke, *Attitudes Toward History* (Berkeley: University of California Press, 1937): 150. Emphasis added.
47. Ibid., 150.
48. Ibid., 411.
49. Wells et al., "Introduction," 4. Embedded quotation from Diane Keeling's insightful article, "Of Turning and Tropes," *Review of Communication* 16, no. 4 (2016): 327.
50. Edbauer (Rice), "Unframing Models of Public Distribution," 5, 13. Emphasis added.
51. Wells et al., "Introduction," 15.
52. *The Ecological Society of America*, "About Page," www.esa.org/esa/about.
53. Gregory J. Seigworth and Melissa Gregg, "An Inventory of Shimmers," *The Affect Theory Reader*, eds. Gregory J. Seigworth and Melissa Gregg (Durham: Duke University, 2010): 3.
54. Jane Bennett, *Vibrant Matter: A Political Ecology of Things* (Durham: Duke University Press, 2010): 116–117.
55. Diane Davis, *Inessential Solidarity: Rhetoric and Foreigner Relations* (Pittsburgh: University of Pittsburgh Press, 2010): 166.

56. Nathan Stormer and Bridie McGreavy, “Thinking Ecologically About Rhetoric’s Ontology: Capacity, Vulnerability, and Resilience,” *Philosophy and Rhetoric* 50, no. 1 (2017): 3.
57. Druschke, “A Trophic Future for Rhetorical Ecologies.” Emphasis removed.
58. See Margaret Graver, *Stoicism and Emotion* (Chicago: University of Chicago Press, 2007): 70–75; Martha Nussbaum, *Anger and Forgiveness* (New York: Oxford University Press, 2016): 261.
59. Anita Girvan, *Carbon Footprints as Cultural-Ecological Metaphors* (New York: Routledge, 2017): 9.
60. See Gregory Bratman, Paul Hamilton, and Gretchen C. Daily, “The Impacts of Nature Experience on Human Cognitive Function and Mental Health,” *Annals of the New York Academy of Sciences* 12491, no. 1 (2012): 118–136.
61. Kay Milton, *Loving Nature: Towards an Ecology of Emotion* (London: Routledge, 2002): 4. Catherine Lutz’s study, *Unnatural Emotions: Everyday Sentiments on a Micronesian Atoll and Their Challenge to Western Theory* (Chicago: University of Chicago Press, 1988), greatly advanced the study of emotion by deconstructing the Western, colonialist attitude that configures emotions as primitive, precultural, and universal, and therefore not subject to cultural variation, but only management through the civilizing forces of rationality.
62. Ibid., 4.
63. Ibid., 4.
64. Ibid., 16.
65. Glenn Albrecht, “Solastalgia and the Creation of New Ways of Living,” *Nature and Culture: Rebuilding Lost Connections*, eds. Sarah Pilgrim and Jules N. Pretty (London: Taylor and Francis, 2010): 217.
66. Ibid.
67. Sara Ahmed, *The Cultural Politics of Emotion* (New York: Routledge, 2004): 8.
68. Daniel Gross, *The Secret History of Emotion: From Aristotle’s Rhetoric to Modern Brain Science* (Chicago: University of Chicago Press, 2006): 8.
69. Ahmed, *Cultural Politics*, 56.
70. Anders Hansen and Robert Cox, “Introduction,” *The Routledge Handbook of Environment and Communication* (New York: Routledge, 2015): 3.
71. Robert Cox, “Nature’s “Crisis Disciplines”: Does Environmental Communication Have an Ethical Duty?” *Environmental Communication* 23 (2007): 15.
72. Ibid., 18.
73. Senja Post, Katharina Kleinen-von Königslöw, and Mike S. Schäfer, “Between Guilt and Obligation: Debating the Responsibility for Climate

Change and Climate Politics in the Media," *Environmental Communication* (2018). Curiously, this sole article from *Environmental Communication* to include "guilt" in the title, does not use the word at any other point.

74. Janet Swim and Brittany Bloodhart, "Admonishment and Praise: Interpersonal Mechanisms for Promoting Proenvironmental Behavior," *Ecopsychology* 5, no. 1 (2013): 432.
75. Ibid., 465.
76. See Robyn Mallett, "Eco-guilt Motivates Eco-friendly Behavior," *Ecopsychology* 4, no. 3 (2012); Mark Ferguson and Nyla Branscombe, "Collective Guilt Mediates the Effect of Belief About Global Warming on Willingness to Engage in Mitigation Behavior," *Journal of Environmental Psychology* 30 (2010).
77. Clare Dannenberg, Bernice L. Hausman, Heidi Y. Lawrence, and Katrina M. Powell, "The Moral Appeal of Environmental Discourses: The Implication of Ethical Rhetorics," *Environmental Communication* 6, no. 2 (2012).
78. Ibid., 223.
79. Miyase Christensen, Anna Åberg, Susanna Lidström, and Katarina Larsen, "Environmental Themes in Popular Narratives," editorial essay, *Environmental Communication* 12, no. 1 (2018): 6.
80. Ibid.
81. Alex Lockwood, "Graphs of Grief and Other Green Feelings: The Uses of Affect in the Study of Environmental Communication," *Environmental Communication* 10, no. 6 (2016): 737, 738.
82. Ibid., 738–739.
83. Ibid., 745.
84. Ann Cvetkovich, *Depression: A Public Feeling* (Durham: Duke University Press, 2012): 11.
85. Ibid., 5, 11.
86. Lockwood, "Graphs of Grief," 741.
87. Lauren Berlant, "Compassion (and Withholding)," *Compassion: The Culture and Politics of an Emotion*, ed. Lauren Berlant (New York: Routledge, 2004); Ann Cvetkovich, *Depression: A Public Feeling* (Durham: Duke University Press, 2012); Sara Ahmed, "The Affective Politics of Fear," *The Cultural Politics of Emotion* (New York: Routledge, 2004); Deborah Gould, *Moving Politics: Emotion and ACT UP's Fight Against AIDS* (Chicago: University of Chicago Press, 2009); Lauren Berlant, *Cruel Optimism* (Durham: Duke University Press, 2011); Sara Ahmed, *The Promise of Happiness* (Durham: Duke University Press, 2010).

CHAPTER 2

Guilt's Plasticity

Abstract This chapter examines three cases of environmental communication to illuminate key rhetorical challenges for assigning, experiencing, and representing guilt for ongoing environmental devastation. I argue that analysis of a particular species of guilt—collective guilt—is essential for understanding the communication of complex environmental issues like plastics pollution. Collective guilt exhibits protean characteristics, which poses challenges for advocacy rhetorics aiming to mobilize collective action. I highlight the critical dividing line between guilt (culpability for an offense) and shame (evaluation of intrinsic inadequacy), as the affective dynamics of each are apt to motivate decidedly different behaviors.

Keywords Guilt · Collective guilt · Plastics pollution · Shame · Blue Planet · Responsibility · Visual enthymeme

Roughly three thousand miles from the Oregon coast sits the Midway Atoll, some 1500 acres of land surrounded by vast expanses of Pacific blue. Its nearest neighbor, Kaua'i, is over a thousand miles away.

Spied from on high, the Midway Atoll appears as a turquoise gemstone suspended in cobalt—a lagoon within an ocean, the result of coral reefs forming stony skeletons atop ridges of long-ago-exploded volcanoes. Walls of coral polyps provide protection and support for marine

T. Jensen, *Ecologies of Guilt in Environmental Rhetorics*,
Palgrave Studies in Media and Environmental Communication,
https://doi.org/10.1007/978-3-030-05651-3_2

species, and the prismatic green water teems with life. The abundant fish and unique location of the Midway Atoll make it attractive to the albatross, one of the world's largest flying birds. The sea-faring and symbolic albatross has found the area to be a fine breeding ground for millennia, well before World War II's Battle of Midway brought it name-recognition to American discourse.

Although the islands of Midway support 17 different seabird species, albatross dominate in overall numbers. Nearly 70% of the world's population of Laysan albatross make Midway their breeding grounds, and have done so for generations. Wisdom, a Laysan albatross and world's oldest known bird in the wild, was banded there on December 10, 1956. In early spring of 2019, at the approximate age of 68 years old, Wisdom successfully hatched yet another chick.

Like all baby albatrosses, Wisdom's latest hatchling must have consistent feedings of regurgitated seafood to survive. Parent albatrosses dutifully trade off responsibilities, one keeping watch while the other forages the sea, scooping up food, and returning home to feed the young. Among the regurgitated squid and krill is plastic: lighters, bottle caps, pen casings. These larger pieces are accompanied by a toxic confetti of microplastics, each bit smaller than a grain of rice. The young albatross receive all of these plastics along with the broken-down seafood, swallowing cephalopods and golfballs, crustaceans and cigarette butts.

If you walk around Midway and see a small collection of plastic grouped in a shallow mound, it's likely where an albatross chick died. After its downy feathers degrade and its carcass is worn away by weather, the pile of plastic it carried in its stomach remains. The islands of Midway are flecked with these macabre monuments: a colorful panoply of stubborn plastic encircled by sun-bleached bones and gray feathers half-buried by the white sand.

A large percentage of chicks die on Midway each year from asphyxiation, choking on plastic lodged in their airways. Others perish from general gut disruption. Many die in a perverse form of starvation: chunks of plastic reduce the amount of available space in their stomachs. Full on plastic, they can't fit in enough food to survive. The physical mechanism that would allow albatross chicks to disgorge isn't developed until they're about to leave the nest, some five to six months after hatching. As their bodies transform in preparation for heading out to sea, fledging albatross learn to cast a bolus—a collection of indigestible elements, like fish bones and squid beaks and increasingly, plastic detritus.

Many albatrosses won't make it to this stage. Life as a newborn seabird has always been hard, but it's much harder now with plastics pollution. Because of plastic, a full third of Laysan chicks will die before reaching adolescence.

Rhetorical Challenges of Ecological Scaling

Communicating a scenario like that of Midway is fraught with rhetorical choices. The sheer enormity of the situation, the ambiguities of culpability and responsibility, and the tragic death of innocent life—all of these factors combine to make this site of environmental communication a formidable challenge. For rhetorics aiming to galvanize action, the conspicuous lack of near-term solutions makes matters more difficult still. This chapter examines three cases that engage the pressing issue of plastics pollution in the oceans. In doing so, they each wrestle with how to communicate *scale*, a central dilemma for all ecological rhetorics, though particularly those focused on global crises.

Each case study invites audiences to grapple with complex questions of culpability—who is at fault and to what extent? Concomitantly, they raise issues of responsibility—the expectation to and capacity for addressing and righting wrongs. In varying ways, we are asked to situate our individual culpability within an ecological crisis. Our crisis is collectively created, without question, but unevenly so, with some actors who are far more culpable, yet who feel less responsible, and vice versa. While making a link between the straw in our cup and the plight of albatrosses may be easily accomplished in the abstract, that relational link is transformed once imbued with affect. It is rendered vastly more complicated when considering options for redressing the wrong. A dense ecology of guilt is at work here, a tangle of deliberations that invite ethical and emotional engagement.

Performing a comparative analysis across multiple modes that engage the same topic yields insight into the challenges of assigning, experiencing, and representing guilt for environmental devastation on such a massive scale. What we see is that guilt exhibits protean qualities—it is conceptually flexible, capable of being molded to varying cultural mores at different historical junctures. The forms it assumes at the individual level are reshaped when scaled to the collective. Guilt's plasticity makes it a rhetorically potent force, but without frameworks that acknowledge this and adapt to it, guilt's influence on beliefs and behaviors will be obscured or remain undetected.

To better perceive opportunities for effective action, we must better understand the rhetorical obstacles that impede current efforts. In their recent assessment of environmental communication's core challenges, DiCaglio, Barlow, and Johnson spotlight scale, characterizing it as a "fundamental rhetorical difficulty of thinking in terms of ecology."[1] They note that "much of the frustration and confusion around environmental degradation arises from the fact that environmental effects are most easily discerned on a larger scale and yet each of us individually do not and cannot act on that scale."[2] Navigating relationships of scale—between individual action and collective consequences, between our personal use of plastic and its environmental ramifications, and between feeling guilty and feeling helpless—are complicated, critical affairs. Ecological attunement facilitates seeing connections between different biophysical systems, as well as across various strata within those systems. Our emotional literacies are often unable to match such dynamism. DiCaglio et al. recommend more attention paid to positive connections, as "scalar relationships … can inspire fear and negative reactions because they reveal connections that were not seen before."[3] Achieving this affective rebalancing will be as difficult as it is necessary.

In his review of Timothy Morton's oeuvre, Joshua Trey Barnett also highlights the centrality of scale, particularly for Morton's concept of *hyperobjects*—things massively distributed across "spatial and temporal scales that cannot be perceived in any usual sense….Think global warming. Think plutonium. Think plastic. Think Styrofoam. Think Earth, too."[4] If your blood pressure spiked from reading that staccato series of environmental issues, the challenge of moving onto more positive affective terrain is evident. Ecological thinking expands opportunities for feeling environmental guilt, the rhetorical dynamics of which can be catalytic or paralytic. "Ecological awareness is going to feel weird in the Anthropocene," Barnett deadpans.[5] In analyzing cases that engage the scalar relationship between environmental crisis and guilt, this chapter clarifies salient factors in communicating collective guilt. The goal is not to propose any one "right" way to assign and represent environmental guilt, but rather elucidate the problems encountered in trying to do so.

In the first case analysis, I explore the prevalent use of numerical data to convey the scale of ecological damage. How, I ask, does this mode function as an affective rhetoric, despite it appearing otherwise? Why is it so common, given its limited capacity to incite engagement? The second case considers the popular television series

Blue Planet II and specifically, viewers' tweets in reaction to plastics pollution in the ocean and their feelings of complicity. As an archive of feeling, the hashtag #BluePlanet2 reveals how viewers are brought to a phenomenological threshold, where distinguishing between guilt (culpability for an offense) and shame (evaluation of intrinsic inadequacy) is perplexing. The interpretation is crucial, however, as guilt and shame orient subsequent action in opposing ways. In the final section, I contemplate a photograph by artist Chris Jordan featuring an albatross who perished from ingesting plastics. The photograph functions as a visual enthymeme, inviting viewers to situate themselves in relation to multiple systems of scale. Before moving into these analyses, it's necessary to first identify the particular species of guilt these rhetorics—and all contemporary environmental rhetorics—are engaging. Whether it is global warming, plastics in the ocean, or mass extinction, the sheer scope of harm transforms how guilt orients us to the problem and toward making amends.

Environmental Guilt Is Collective Guilt

"There is something distinctively modern in our conceptualizing guilt as a property of individuals rather than collectives," political philosopher Samantha Ashenden writes.[6] In contrast, for many societies in the course of human history, she notes, "guilt was both collective and objective."[7] The contemporary approach to guilt in fields from psychology to jurisprudence emphasizes personal, direct actions—an individual either doing wrong or failing to do right, with more focus on the former and increasingly less on the latter. This conception is the product of cultural, legal, religious, and philosophical rhetorics that, over centuries, have shaped the emotional literacies and vocabularies that aide in the identification, interpretation, and expression of certain patterns of feeling that we call guilt. Just as we are subject to guilt's persuasive capacities, the concept of guilt is itself subject to rhetorical forces.

Guilt's alignment with the individual citizen is but one manifestation of a broader, protracted cultural shift toward individualism. As a body of thought, for example, classical liberalism has left a lasting impression on the formation of what Lauren Berlant calls our "National Symbolic"—a cluster of texts that help construct national public discussions.[8] The philosophical traditions of idealism and individualism converge in liberalism, engendering a view of society that privileges the sovereignty of the rational, free-willed individual at its foundation. Within the context of the European Enlightenment, the individual is defined by a (supposed) autonomy and innate capacity for pure reason, which, as Lakoff reminds us, is characterized as "conscious, unemotional, logical, abstract [and] universal."[9] Coexistent with this Enlightenment conception of the rational, individual subject arose a construction of guilt in legal and moral realms that stresses personal responsibility and intentional, empirical actions of wrongdoing.

One consequence is that other formulations of guilt are cast into the shadows. *Collective guilt*—feelings of distress and remorse for being part of a group that has committed or is committing unjust harm[10]—becomes particularly less legible as a form of guilt. Collective guilt emerges from active or passive participation in a group or system (e.g., governmental, economic, organizational, etc.) that transgresses some ethical boundary. Consider, for example, that if you pay taxes in the United States, you are helping fund military drone-strikes that, in 2017 alone, killed more than 6000 civilians in Iraq and Syria. You needn't be directly involved in causing those casualties to experience some degree of guilt for them, you need only to recognize your financial support of the system that carries out the inadvertent murder of innocent people. This is not an accusation, it is a statement of facts. To truly ruminate on its implications, however, is to consider the emotional logic and potency of collective guilt. The affective resonances and cognitive dissonance this example may have elicited are worthy of reflection, as these sensations and quick cognitive maneuvers are insights into collective guilt's unique constitution.

Even if you did not pay taxes, it is still possible to experience a degree of collective guilt, since identifying with the group is enough to generate a sense of culpability. "[T]here is considerable agreement" in the field of social psychology, where the contemporary study of collective guilt is arguably most developed, "that a shared social identity with the perpetrator group is essential for eliciting collective guilt."[11] In other words, the experience of collective guilt hinges significantly on one's identification with the group. "Feeling guilt for events that an individual is not personally responsible for is possible because people can and do categorize themselves as members of a group," note social psychologists Nyla Branscombe and Bertjan Doosje.[12] Just as we are capable of collective pride when, for example, we celebrate the wins of our chosen teams in sports, so too are we capable of collective guilt.

The everyday production and analysis of collective guilt is critically important to environmental communication for a simple reason: environmental guilt *is* collective guilt. Though the disquieting sensations of guilt may be prompted by personal action, perhaps a mild twinge when idling in a car, for instance, that guilt is associated with the collective harm it contributes to. When we feel bad about a negative impact we're making or a positive impact we're failing to make, it is inextricably bound up in the awareness of participating in something much larger. An ecological perspective draws connections between actions and their impacts on biophysical ecologies, bridging the miniscule emissions of a quick car trip and the cumulative effects of burning fossil fuels, for example. Although the link between individual act and collective harm, between perpetrator and offense, is extended conceptually, it can nevertheless be felt intensely. Statistically, one's daily commute is negligible in the broader picture of global carbon output. Ecologically, there are numerous steps between driving and its impact on, among other things, thermal expansion of ocean water. Emotionally, however, the distance between starting up a combustion engine and rising sea levels can collapse quite easily. Such is the plasticity of guilt. Its flexibility of form allows it to range across scales of breadth and depth with both ease and speed.

Because our modern conception of guilt is heavily premised upon the individual and their intentional actions, however, environmental guilt is conceptually and phenomenologically obscured. As a unique form of collective guilt, its rhetorical dynamics largely elude the emotional literacies we have developed for describing, expressing, and responding to guilt. "This assumption of individualism sustains the eschewal of collective guilt," Ashenden affirms, and yet, she adds, the concept nevertheless persists: "collective guilt is a notion that modern political reason cannot embrace and yet which it cannot entirely disavow."[13] As a result, collective guilt occupies a liminal zone—many people feel it, and simultaneously, have impoverished frameworks for identifying and contemplating it.

To date, collective guilt has yet to receive sustained attention within the field of environmental communication. There are, however, several studies within environmental psychology that have examined collective guilt to determine how it functions in relation to individual behavior. One study finds, for example, that "when college students bring to mind their standards for environmental behavior, ranging from recycling to buying a fuel-efficient car, doing so creates eco-guilt, which motivates efforts to protect the environment," leading the author to conclude that "eco-guilt predicts eco-friendly behavioral intentions [and] pro-environmental attitudes."[14] This chapter argues that environmental guilt is more complicated—and less predictable—than this study suggests. The following chapter, "Eco-friendly Scapegoats," reveals how "pro-environmental attitudes" have complex, and at times contradictory dynamics, and thus benefit from a rhetorical lens for analysis.

Although research on collective guilt in environmental psychology offers valuable insights, I contend that the study of environmental guilt would benefit from a profusion of methodologies, especially those from the humanities. A rhetorical perspective is particularly germane, given that collective guilt functions through one's identification with a group. Examining the suasive forces that bring that identification into being, that shape and direct it, are critical for understanding how environmental guilt orients attention and action. Any feelings of guilt we might experience in connection to the devastating situation occurring at Midway, for instance, are complexly layered and reticulate in ways that are difficult for quantitative approaches to capture. Feelings of guilt for using plastic can function quite differently than other, more recognizable forms of guilt, because plastics' creation is massively distributed, yet the material is intimately tied to the routines and objects of everyday life. Rhetorical theory, in conjunction with affect theory and critical emotion studies, can help illuminate these relationships.

Plastic's Material Persuasion

The ubiquity of plastic makes it a worthy material through which to analyze environmental rhetorics. Plastic, after all, is quite persuasive. As a material, it has shaped the modern economy because it can be molded into a staggering variety of forms. Plastic, as a result, is pervasive: a plastic bag was recently discovered at the bottom of the Mariana Trench, the deepest part of the ocean, some seven miles below the surface.[15]

An ecological perspective animates and accelerates the process of scaling out, of shuttling between the specific and its system, as well as across different systems of scale. It encourages you to draw connections between the jumble of plastic bags stashed under your sink and the plastic bag suspended in the deepest, darkest trough of the ocean floor. The predilection of an ecological orientation is to tease out these interrelations, and examine how physical matter intertwines with matters of concern. In a bitter twist, plastic actually helps to illuminate ecological relationships by virtue of its durability, allowing itself to be traced as it passes from system to system, from a river stream to a human bloodstream, from plankton to platelet. "While the smallest or simplest body or bit may indeed express a vita impetus, conatus or *clinamen*," Jane Bennett writes regarding vital materialities, "an actant never really acts alone. Its efficacy or agency always depends on collaboration, cooperation, or interactive interference of many bodies and forces."[16] Exploring plastics pollution rhetoric from an ecological orientation, it follows, entails consideration of how plastics' materiality functions as a persuasive agent, but always in relation to other elements.

Take the albatross, for example, whose plight with plastic is that of nearly all seabirds. A recent research study published in the *Proceedings of the National Academy of Sciences* estimates that 90% of seabirds have ingested plastic, up from the estimated 29% in 1962.[17] Plastics are one factor—albeit a very significant one—in the staggering decline of seabird populations, whose numbers have dropped nearly 70% since 1950, the equivalent of 230 million deaths.[18]

Such statistics of ecological upheaval are poised to get worse before they get better. At the current pace of plastic production, it's estimated that 99% of all seabird species will be affected by 2050.[19] By that point, there will be, pound for pound, more plastic in the ocean than fish. We may reach that point sooner, however, as marine populations are plummeting faster than previously anticipated, while plastic production is increasing with astonishing velocity. The current doubling rate of plastic averages 11 years.[20] That is, the amount of new plastic to be produced over the next 11 years will be equivalent to *all* the plastic produced prior—*ever*.

Precisely because plastic is so abundant and familiar, we are largely desensitized to its presence. Although we may attune to the ubiquity of macroplastics in our everyday lives—in plastic bottles, for example—the near-invisible presence of *micro*plastics (smaller than a grain of rice) and *nano*plastics (1000 times smaller than an algal cell) is growing rapidly. Since the majority of car tires are now made from synthetic rubber, microplastics get shaved off when we drive, then make their way into our watersheds. These plastics are all but impossible to capture, let alone recycle. Similarly, clothing is rarely identified as a major source of plastic pollution, despite the increasing prevalence of synthetic fibers and fabrics. With each washing of these garments, hundreds of thousands of polyester and acrylic microfibers escape with the flushed water. A 2017 report found that up to 30% of the plastic in our oceans comes from tire dust and textile fibers.[21] This number is bound to increase, because polyester overtook cotton and wool in 2007 as the world's principal fiber of choice. It has only expanded its market share since: between 2007 and 2017, more than twice as much polyester has been produced than cotton.[22]

Even if you strive to eliminate plastic from your life—using organic cotton sacks to transport bulk groceries in glass containers on a bike made of bamboo—it's nevertheless highly probable that microplastics and their primary compounds are in your body. In fact, you're likely drinking plastics on a regular basis. An international study of tap water found that the vast majority was contaminated with microscopic plastic fibers—83% of all samples collected across six continents.[23] The United States had the highest rate of concentration, with 94% of samples containing plastic fiber. A study by the Center for Disease Control and Prevention found that 92% of the more than 2500 tested subjects had measurable levels of bisphenol A—more commonly known as BPA—a principal compound of the most widely distributed plastics.[24] Even a few parts-per-million have been shown to disrupt endocrine systems, negatively affecting reproductive and immune health.[25]

The omnipresence of plastic in our stores and seas, soil and cells means that we are participants in a complex, complicated problem. *Complicated* not only by virtue of the sheer scale of plastic pollution and its impact on multiple ecosystems in multiple ways but also by plastic's convenience in our everyday lives. *Complex* in that the situation involves ever-evolving ecological relationships and numerous unknowns regarding the impact of a man-made material designed to last more than one thousand years. Plastics don't just surround us, *they permeate us*. In doing so, they illuminate ecological connectivity in a very tangible, toxic way.

Data Dumping as Affective Rhetoric

Perhaps the previous section had a familiar feel to it, even if the content was unfamiliar, as the style is prevalent within environmental communication. It is an exemplification of what Timothy Morton calls "an *information dump*."[26] A brief review of journalism reporting on plastics pollution will yield numerous variations on the mode, such as this paragraph plucked from a recent article:

> We use more than 300 million tons of new plastic every year. Half of this we use just once and usually for less than 12 minutes. Plastic production has increased 20-fold since 1964, and is expected to double again in the next 20 years and almost quadruple by 2050.[27]

If that last batch of statistics from the newspaper felt less poignant than the data delivered a few pages back, it would not be surprising. Repeated use of numerical data to communicate matters of scale often yields a sense of distance and eventually, numbness. For many, these last few statistics, which are utterly horrifying, are likely to yield little more than an anesthetized "damn, that's a lot of plastic" or a somewhat deflated feeling of "we're screwed."

As a prominent mode of communicating environmental issues—one that is particularly abundant in discourse about plastics—it is worth examining how this style of numerical info-dumping functions rhetorically. A key challenge for contemporary environmental communication, we've established, is in *scaling out*—conveying the cumulative consequences of many acts in a compelling manner, such that others understand the raw immensity of the situation, yet not become so overwhelmed that it leads to acedia or cynicism. It is at this rhetorical junction that one commonly encounters a turn to the numerical. Here, information about a global environmental issue is imparted through the piling up of statistics and scientifically derived data. While the prevalence of the data-dump is reason enough to merit study, I argue that it is imperative to inspect how a style that forefronts facts operates on an affective and emotional level.

One reason for attending to the affective register of this "eco-data dump" style is simple, yet has serious consequences. As cognitive psychologists Paul Slovic and Daniel Västfjäll put it, "Many people do not understand large numbers."[28] A substantial body of scientific and humanistic research strongly suggests that "large numbers have been found to lack meaning and to be underweighted in decisions unless they convey affect (feeling)."[29] A fact like *we use 300 million tons of new plastic each year* strains the limits of imagination in trying to conceptualize what 300 million tons looks like. Trying to envision the 23 billion plastic bags New York state uses each year being tied together and stretched to the moon and back—more than *13 times*—with any remote degree of accuracy is simply inaccessible to us, utterly unattainable.[30] These facts are, in the most accurate sense, *trivia*, floating about without the narrative context to animate and sustain action.

Why, then, given the limitations of comprehension, does the listing of numerical facts to emphasize severity remain a prevalent feature within dominant genres of environmental communication? The editors of *Numbers and Nerves: Information, Emotion, and Meaning in a World of Data*, offer one credible reason: "We've come to believe that truth inheres in numbers, and people who speak (and write) the language of numbers *appear* to know what's going on in the world."[31] The pursuit of credibility, by both authors and audiences, and the desire for trustworthy information is, understandably, a consequential factor. "Many people in industrialized societies accept without question the special form of veracity that seems to attach itself to numbers," Scott Slovic and Paul Slovic write, and yet, they note, "this [acceptance] is the result of cultural determination, not an inevitable and permanent truth of human nature."[32] The ecological data-dump, in other words, is a byproduct of western Enlightenment values and reverberates with its assumptions. It echoes a framework of credibility that privileges scientific rationality and its numerical idiom, in part because they are assumed to provide an unemotional—and therefore unbiased, and therefore true—account of reality. The same cultural rhetorics that have shaped our individualistic conception of guilt have also helped imbue numerical information with the aura of authenticity.

But western Enlightenment's enduring legacy pitting reason contra emotion is only one element of the data-dump's rhetorical formation. The style is routinely employed in ways that extend well beyond the ostensible purpose of imparting information. Its implementation within environmental rhetorics of decline and devastation often suggests, if not insists, that the audience *feel* a certain way in response to a stack of statistics. How do these numerical heaps function as *pathema*—evidence presented as cause for a feeling—and how, subsequently, do those feelings then serve as evidence for taking action on the issue of plastics pollution?

Morton argues that the "ecological information delivery mode" has a pugnacious quality: "It seems to be shouting at us—*Look. I'm a fact. You can't ignore me.*"[33] The confrontational manner fuels a sense of urgency, Morton writes: "it's as if it is screaming, 'Look, can't you see? Wake up! Do something!'" The information overload aims to be both the catalyst and justification for a change in behavior. The underlying logic is unambiguous: deploying facts that disclose ecological harm confronts the audience with a scene of injustice, and in doing so, enjoins one to contend with that injustice in some capacity. At root, the operating assumption is that once you *know* about an ecological harm, you are entangled into an ethical relation that obligates you to *act*.

We should read the confrontational quality that Morton describes, then, not necessarily as a tonal aesthetic, but rather, as an expression of the mode's constitutional logic. The ecological data-dump confronts you with the obligation *bear witness*—to be present to the sufferings and injustices of environmental devastation. When combined with the authoritative aura of statistics, the data-dump invites a narrow set of possible feelings. As Morton puts it, "the delivery mode seems to be saying *Don't question this*, or even *You should feel very bad if you question this.*"[34] Even if you do not question the facts, however, the mode still seems to be saying *You should feel very bad about this.* You are not only a witness to harm, after all, you are also the perpetrator. As a rhetorical mode of environmental communication, the data-dump exhibits a patina of objective knowledge, but its deployment is more often than not an exhortation to *feel*, and to feel in a manner that leads to active engagement. The information overload is utilized principally as an affective rhetoric. But poorly.

Our cognitive limitation for digesting large numbers plays a part in this failure, as does the seemingly undiminished attachment to Enlightenment ideals that posit rationality and emotion in an inverse relationship, such that the more one increases the other decreases. Another factor to consider is our propensity for becoming inured to statistics of horror, a process known as "psychic numbing." As Slovic and Västfjäll point out, "psychophysical research indicates [that] constant increases in the magnitude of a stimulus typically evoke smaller and smaller changes in response [resulting in] our inability to appreciate losses of life as they become larger."[35] Affectively, this manifests as numbed detachment and even casual callousness. Exposure to a litany of ecological facts commonly results in feelings of discouragement, melancholy, anxiety, regret, shame, and, of course, guilt.

Indeed, for Timothy Morton, the principal shortcoming of this style's popular usage is its propagation of guilt. "We yell facts thinking that this will inspire and persuade people,"[36] he laments, but the outcome is otherwise: "An awful lot of stuff from the ecology world … is about making people feel really, really guilty all of the time."[37] Morton perceives the strong link between collective inaction and pervasive feelings of guilt for global warning and its coeval of mass extinction. His solution: do away with the guilt. The proposal merits quoting at length:

> As you move through this book [*Being Ecological*], watch any feelings of guilt that come up. After all, guilt is scaled to individuals. But individuals are *in no sense guilty* for global warming. That's right—you can totally let yourself off the hook, because starting the internal combustion engine of your car every day is statistically meaningless when it comes to global warming. Big corporations are obviously capable of having this effect. But their employees' effect is, to use the phrase again, statistically meaningless. Several thousand years from now, nothing about you as an individual will matter. But what you did will have huge consequences. This is the paradox of the ecological age. And it is why action to change global warming must be massive and collective.[38]

Morton correctly claims that guilt is scaled to the individual. The reasons for that alignment, however, are critical for understanding why environmental guilt is so difficult to discuss and digest. Guilt's plasticity lent itself to being molded over time in accord with cultural principles and institutions. As guilt became closely associated with the individual, conceptions of collective guilt withered away in the emotional literacies that circulate through cultural pedagogies. The emotional frameworks used to make sense of guilty feelings increasingly reflected only the individual, and thereby deflected recognition of collective guilt—a species of guilt whose uniqueness and value lay in its power to illuminate the connective threads that bind us into groups with shared responsibilities. Guilt is scaled to the individual, yes, but rhetorically—not essentially. Historically, it has not always been so tightly tied to the individual. The rhetorical plasticity of guilt means that it needn't remain so in the future, either.

The "paradox of the ecological age" Morton references is simultaneously an ethical, phenomenological, rhetorical, and material paradox. That is, Morton may mount a viable argument in the realm of ethics for why one individual is "in no sense guilty" for the totality of global warming. Aided by a modern conception of guilt, he might argue that one can only be held accountable for their direct, individual contributions to global warming, which, because they are statistically negligible, offer an almost full exoneration from the charge. But this argument, paradoxically, does not hold sway in an ethical realm that acknowledges collective guilt, which governs by a different calculus of blame.

Nor does this ethical argument necessarily have any jurisdiction over the phenomenological realm. Telling someone that they are "in no sense guilty" for global warming dangerously discredits the *sensorial* experience of guilt. In contrast, a rhetorical approach to environmental guilt recognizes that feelings of guilt do not neatly correlate to cultural and juridical conceptions of what is and is not deemed guilt. The relationship between feeling guilty and "being" guilty is highly elastic, as the metrics used in determining each are at times quite dissimilar. Whether or not your actions are statistically meaningless may, in short, be meaningless. The limitations of numerical figuration to communicate ecological issues of scale apply to attempts at absolution, too. As I demonstrate in the following section, however, the phenomenological qualities of environmental guilt present their own set of challenges.

#BluePlanet2

In the final episode of *Blue Planet II*, a 2017 BBC nature documentary on ocean life, viewers are made witness to the impact of plastics on marine life. In one sorrowful scene, a short-finned whale is shown carrying her young in her mouth, which she had done for days, despite the fact the calf is dead. We're told the death was possibly caused by high toxicity levels in the mother's milk as a result of plastic ingestion.

For David Attenborough, famed naturalist and narrator, another incident was even "more heart-rending" than this portrait of a grieving mother, more painful than any other shot in the entire series.[39] Writing in a blog post following the show's release, Attenborough describes the scene: "In it, as snowflakes settle on the ground, a baby albatross lies dead, its stomach pierced by a plastic toothpick fed to it by its own mother, having mistaken it for healthy food."[40] In the last minutes of the season finale, Attenborough, striding along a beach, delivers these concluding remarks:

> [N]ever before have we been so aware of what we are doing to our planet—and never before have we had such power to do something about it. Surely we have a responsibility to care for the planet on which we live? The future of humanity, and indeed of all life on Earth, now depends on us doing so.[41]

In three sentences, Attenborough's exhortation exemplifies a distilled environmental rhetoric of guilt. Its enthymematic delivery of premises is concentrated, its articulation of the stakes unapologetically expansive, and its challenge to the audience—at once a question and a strident argument—makes for a dramatic moment.

Blue Planet II's spellbinding shots of marine life, its beloved host, the vignettes of ecological harm, and its concluding assignation of blame and corresponding invitation to take responsibility, all helped generate a wave of public response in the wake of its final episode. Following the finale's airing, which first ran in the United Kingdom before being released internationally, viewers took to Twitter to comment, quip, and confess. The conversation surrounding the show had already coalesced around the hashtag #BluePlanet2 by that point, and like all other hashtags, #BluePlanet2 signaled the conversation users were participating in, aggregated the voices engaged on that topic, and facilitated that conversation by acting as a search tool. The thread is rich with explicit and implicit articulations of feeling—grief, joy, frustration, wonder, guilt, skepticism, anger, disbelief, and shame, among others.

The public statements curated through #BluePlanet2 constitute an archive of ecological expression. This compilation also serves as an archive of feeling—an assemblage of affective and emotional patterns that help constitute publics. Ann Cvetkovich argues in *An Archive of Feelings* for reading such emotional expressions as more than a mere collection of individuals engaging with public culture. She also discourages any analysis that would approach them as the public expression of private affective response. She advocates instead for "a collapsing of these distinctions [between private and public feeling] so that affective life can be seen to pervade public life" and recognized "as the foundation for the formation of public cultures."[42] The sentiments that reverberate across #BluePlanet2, in other words, are not an aggregate of individual feelings, but enunciations of public feelings whose circulation is primary to how publics—like those engaging with plastics pollution—come into being.

My theoretical approach differs slightly from Cvetkovich's, for whom archives are repositories of emotions encoded within texts and their surrounding practices.[43] The ecological orientation I take sees the archive not as a fixed depository, but as a dynamic structure created and maintained through a web of affective, rhetorical, and ecological relations. In this way, I follow Sara Ahmed, for whom the circulation and entanglements of affect, emotion, and rhetoric creates the *impression* of boundaries. As she explains, "emotions are not 'in' either the individual or the social, but produce the very surfaces and boundaries that allow the individual and the social to be delineated as if they are objects."[44] Ahmed's approach evokes ecological overtones, as boundaries are always in the process of being created and maintained through repetition and exchange; they are never fixed nor hermetic.

Approaching the hashtag #BluePlanet2 as a dynamic archive of feelings allows me to trace the contours and patterns of emotional expression in an expressly public venue. While recognizing that linguistic expressions of emotion are never equivalent to the embodied experience, tweets regarding *Blue Planet II*'s portrayal of plastics pollution illustrate the active configuration of public feeling. Regardless of whether the tweets are grasping to describe a deeply felt experience or are enacting what one believes they *should* feel—a performance of emotional expectations—they depict publics actively composing the emotional ecologies of environmental rhetorics. This is a methodology that, in the words of cultural theorist Ben Highmore, "[puts] mood and feeling on the side of *process* rather than on the side of product."[45]

To situate a series of tweets curated through the use of a hashtag is also a conscious choice to highlight a medium that celebrates a reactionary form of experiencing daily life. The vast majority of tweets quickly appear and disappear, ephemeral flashes of everyday life. As Alex Lockwood contends in reviewing how affect and emotion have been studied in environmental communication, "the actual ways in which affect works in and through social systems remains under-theorized and under-employed as a means to further understand how environmental communication can lead to behavioral changes, both positive (pro-environmental) and negative (obstructive)."[46] To bring #BluePlanet2 tweets into analytical focus is to emphasize the mundane ways in which social pedagogies of emotions are formulated. Just as small amounts of carbon emission accrue over time to have profound impacts on collective structures of life, the structuring of emotion occurs through repetition of all-too-ordinary modes of address.

Phenomenological Ambiguity in Collective Guilt

#BluePlanet2 tweets portray a rich mixture of emotional responses with both positive and negative valences, as well as differing levels of intensity (Fig. 2.1). These tweets position feelings of guilt operating in conjunction with other emotions, and notably, with those that aren't typically associated with guilt, such as wonder, joy, and awe. Forms of appreciation and enthrallment complicate the negative valence associated with guilt, affirming the premise that environmental

Fig. 2.1 #BluePlanet2 Tweets showing mixture of emotion

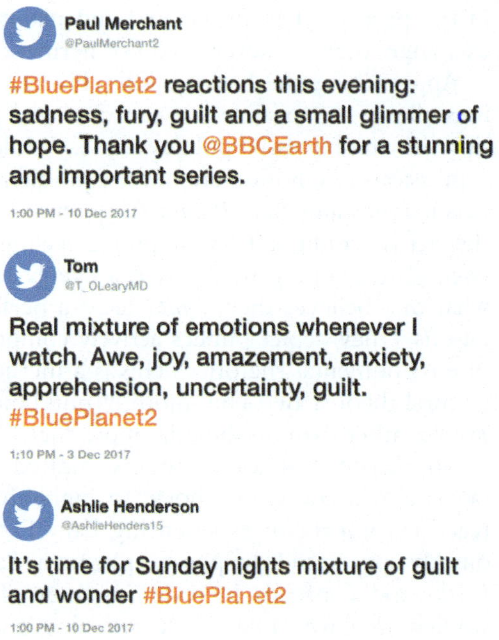

guilt is always a braided emotional experience. They also remind us that even when guilt is foregrounded in discourse or embodied experience, it is always operating within a reticulate system of other emotions.

Unsurprisingly, a dominant strain in #BluePlanet2 tweets expresses a sense of guilt over plastics pollution in the ocean (Fig. 2.2). These three posts are representative of the substantial number of tweets that explicitly identify and label guilt. Unlike the previous set of tweets, which emphasize a mixture of multiple, ostensibly discordant emotions as the principal point, these tweets spotlight guilt as the dominating response. Moreover, they note the quantity or intensity-level of guilt: *a good dose*, *tremendous*, *a lot*. They also gesture to more subtle patterns. The first and third tweet, for example, both particularize the kind of guilt being experienced: *plastic* guilt, *environmental* guilt. The efforts to specify, one by way of reference to the victim (the environment) and the other to

Fig. 2.2 #BluePlanet2 Tweets expressing guilt

the mechanism of harm (plastic), suggest recognition that this form of guilt has unique constitutive elements that characterize the texture of its experience.

Another salient trend is typified by expressions of guilt tied to the human race (Fig. 2.3). Here, guilt is assigned to humans as a collective, either as a biological species or through its abstract essence—"humanity." In the second tweet, a sense of shock is registered from witnessing ecological devastation, yet that destruction is attributed to the human *species*, in contrast to a particular culture, set of specific nations, or collective actions incentivized through economic systems. In the third tweet, the experience of guilt is vividly connected with a sensed incapacity for action—complicity in global ecocide creates a sense of guilt that is *crippling*. And the fourth tweet explicitly brings guilt together with shame. In each, the culpable agent is scaled to a perpetrator with indistinct boundaries—humans, writ large.

The collective element of environmental guilt is rendered explicit in these posts. In collective guilt, feelings of remorse and anguish are experienced in response to unjust harm committed by a group that one

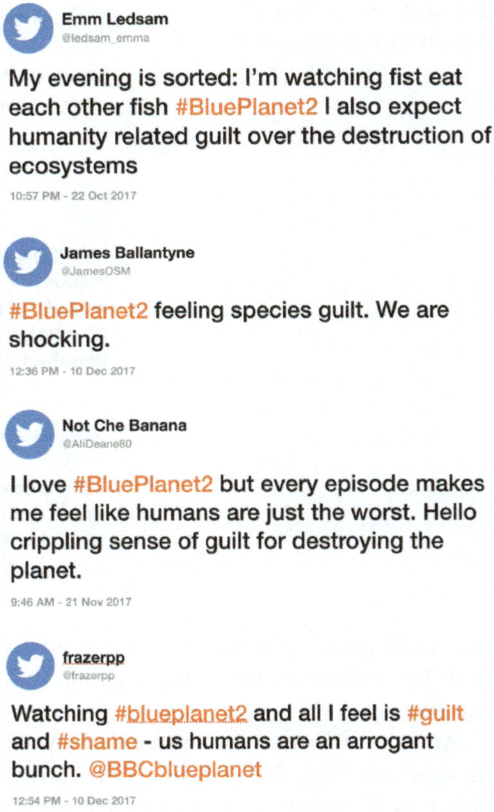

Fig. 2.3 #BluePlanet2 Tweets showing shame and guilt

identifies with. Here, the group is construed as humans *qua* humanity. Following Ahmed's theoretical approach, I see these statements as both description and creation, acts that point out and simultaneously produce the "boundaries that allow the individual and the social to be delineated as if they are objects."[47] In this case, however, the social body being delineated is totalizing in effect. By attributing environmental harm to humans *in toto*, the emotional ecology begins to transform, catalyzing guilt's transmutation into shame.

"Phenomenologically, guilt may *feel* similar to shame but it functions quite differently from shame," philosopher Deborah Tollefsen notes.[48] The ambiguity is amplified further when considering collective guilt and collective shame. Yet the difference between the two is critical for understanding contemporary environmental rhetorics. Guilt operates in relation with moral behavior—via emotional judgments about how one *acts*. Shame operates in relation to character—via emotional judgments about who one *is*. "[P]eople feel guilty for *what they have done*, but feel ashamed for *who they are*," social psychologists Lickel, Schmader, and Barquissau summarize, further noting that, "[while] the distinctions between these emotions might seem subtle, they are supported by several empirical studies."[49]

Though guilt and shame tend to blur together in our physiological registers, distinguishing the source of bad feelings demands attention to nuance and subtleties of experience—is this because of something I *chose to do* or because of who I *am*? The differentiation warrants attention because the two emotions orient action in sharply divergent ways. "Shame causes one to hide, to avoid others, to avoid interaction with those in whose eyes we have been shamed. Guilt, on the other hand, often results in an opening up to others," Tollefsen writes.[50] The contrasting responses can be traced to how each emotion configures one's agential capacity. Shame frames the situation around an intrinsic condition: because the source of the wrong is fixed and cannot be changed, shame tends to fuel avoidance behavior. Guilt focuses attention on specific past actions: because it implies a capacity to have acted differently or act differently in future circumstances, guilt tends to encourage reparative action.[51] The motivational orientation of each emotion is rooted in how they define one's capacity for action. The boundary zone, then, between guilt (culpability for an offense) and shame (evaluation of an intrinsic inadequacy) is critical, as the affective dynamics of each are apt to propel one in decidedly different directions.

Such emotional energy is crucial for mobilizing collective action, particularly when the call to action is as open-ended as Attenborough's in his final address: "Surely we have a responsibility to care for the planet on which we live? The future of humanity, and indeed of all life on Earth, now depends on us doing so."[52] If one *feels* the immensity of plastics pollution, in both its tragic effects and in the lack of obvious solutions, and is then reminded of what is at stake in no less immense terms, combined with a capacious recommendation to take on the "responsibility to care for the planet," it is not difficult to imagine how one might feel overwhelmed. If coded as shame, wherein harm is ascribed to humans as a *species*—in contradistinction to a particular economic system or set of cultural values, for example—then there will be even less motivation to act, for a simple reason: A human can't change the fact they're human. If the experience is coded as guilt, however, the attention will be directed toward specific actions, both individual and collective. These discernments are made more challenging by the dearth of emotional literacies that would help identify and digest our emotional experiences.

Where All Are Guilty, Nobody Is

The information-overload style, it's been shown, is inadequate for communicating the evidence of misconduct and for failing to inspire engagement. *Blue Planet II* inspired many to feel deep concern about the impact of plastics on sea-life, overcoming one critical obstacle of persuasion, but the phenomenological experience of collective guilt presents its own hazards. Discerning between collective guilt and collective shame can be difficult, particularly when the group is large (e.g., the entire human race) and when the harm done is particularly heinous—as in cases of genocide and ecocide, with slaughter occurring on such a massive scale that it's hard to comprehend.

It is perhaps unsurprising, then, that significant deliberation over collective guilt occurred in the aftermath of World War II. The holocaust provoked reflection on the political, moral, and rhetorical mechanisms that enabled the atrocities to occur, and debates waged around questions of complicity and responsibility, as appalling acts could not have happened, without high levels of public acquiescence and passive cooperation alongside active involvement. This section extends analysis of the rhetorical dilemma in assigning collective guilt by briefly charting how philosopher Hannah Arendt grapples with the issue in the wake of WWII. Both in her essays as well as in her own struggle to theoretically resolve the tangled issue, Arendt helps to illuminate the challenges posed by collective guilt.[53]

Arendt elucidates the conceptual difficulties of rendering judgment for collective acts within a framework of guilt anchored in the individual. In fact, she outright rejects the category of collective guilt on those grounds: "Guilt, unlike responsibility, always singles out; it is strictly personal."[54] Arendt is not only concerned with the ethical improprieties and political dangers of classifying an entire demographic as guilty (i.e., the German people), as that form of essentialization, which jettisons guilt's mooring to specific deeds, was used to holistically condemn Jewish people. She is also troubled by how it detracts focus from those who are truly guilty. "[The] cry, 'We are all guilty'… has actually only served to exculpate to a considerable degree those who actually were guilty," she argues, thus reframing the statement as a "declaration of solidarity with the wrongdoers."[55] For Arendt, collective guilt threatens to flatten the varying levels of individual guilt, such that its full application could have an inverse effect: "Where all are guilty, nobody is."[56]

Culpability for Holocaust atrocities, of course, do not fall only on those who performed the physical acts of murder. In her 1963 book, *Eichmann in Jerusalem: A Report on the Banality of Evil*, Arendt analyzes the trial of Adolf Eichmann, a Nazi officer charged with crimes against humanity for his role in organizing logistics for deporting Jewish captives to reservations and concentration camps. Arendt focuses primarily on the legal category of guilt for what she deems a "new type of criminal,"[57] namely, those who argue, as Eichmann did, that they did none of the killing themselves; they were unaware of the slaughter; they were merely following direct orders and national law; they had no desire nor malintent to harm anyone. Eichmann's defense was that he was simply doing his job. Arendt offers fulsome critique of both this defense and the trial itself, though particular attention is given to the role of awareness in assigning guilt. Eichmann was guilty, not of being a monster, she argues, but of not thinking—of not contemplating the consequences of his actions, nor critically questioning his role within the system he was a part of. The "banality" in the book's famous phrase, "the banality of evil," refers to the quotidian, administrative actions that help facilitate mass murder, yet also and critically, the *thoughtlessness* with which they are accomplished.

Arendt draws attention to how social conditions help foster this mindlessness, which is indispensable for the implementation of mass murder. The Nazi regime had strategically "destroyed the neutral zone in which the daily life of human beings is ordinarily lived," she observes, such that every individual "depend[ed] either upon committing crimes or on complicity in crimes."[58] The boundaries between aiding systematic crimes and simply going about everyday life were made porous, at times indistinguishable. "The true problem," she argues, is how, in those conditions, one can render legal and moral guilt for "people among whom the boundaries dividing criminals from normal citizens, the guilty from the innocent, have been so completely effaced."[59] When social conditions are such that indirect participation in mass brutalities is an established norm, how is one to conceptualize individual guilt?

There are three implications of Arendt's arguments that are salient for examining environmental guilt from a rhetorical perspective. The first is that collective guilt has the potential to distract and diffuse attention from those who arguably hold more fault for committing harm. The second implication is an extension of the first. In focusing on Eichmann, Arendt emphasizes that an individual's objective guilt may increase as they move into positions where their job is the *orchestration* of harm (to be carried out by someone else). She quotes the judges on this matter, who declare, "the extent to which any one of the criminals was close or remote from the actual killer of victim means nothing.…On the contrary, in general the degree of responsibility increases as we draw further away from the man who uses the fatal instrument with his own hands."[60] In other words, widespread discourse of collective guilt may work to muddy our evaluation of culpability—especially for those agents who worked to strategically organize wrongdoing to be carried out by others.

The third implication is an emphasis on questioning how one's social conditions place them into positions of unavoidable complicity. As Ashenden writes in her reading of Arendt, "complicity on an enormous scale makes it less likely that individuals will recognize their guilt, but it also erodes the very possibility of recognizing and therefore judging crime per se; atrocity threatens to become elusive through its very ubiquity."[61] While Arendt objected to the category of collective guilt, she was savvy in showing how atrocities occur in part through social conditions that normalize complicity, rendering it more difficult to question. For Arendt, however, one's individual guilt for participating in collective harm is evaluated not only, nor even primarily by their direct contributions, but by the extent to which they question how their cultural, political, and economic institutions oblige acts of participation that, though perhaps mundane, nevertheless facilitate atrocity. Individual guilt increases, she suggests, in relative proportion to how thoughtless one is in committing acts that harm or in failing to act to stop them.

Arendt's analysis adds several layers of complexity to the rhetorical potentials and pitfalls of collective guilt, even as she rejects it as a viable legal or moral concept. In drawing a strong link between collectively created atrocity and societal conditions that normalize complicity, she insists on the need for appraising how one is involuntarily enrolled into acts of systematic destruction. Although she urges us to identify the social mechanisms of complicity, she simultaneously warns us about feelings of collective guilt, which are apt to arise after acknowledging one's contributions to misconduct. The obscuring and exculpatory power of collective guilt has the rhetorical capacity to flatten varying levels of individual guilt, which is its own kind of injustice.

For those who maintain that guilt "always singles out [and is] strictly personal"[62] (as Arendt does) and that "guilt is scaled to individuals"[63] (as Morton does), collective guilt is an unworkable concept. Both Morton and Arendt acknowledge, however, that many people experience feelings of collective guilt. For Morton, these are unnecessary and unproductive feelings because "individuals are *in no sense guilty* for global warming" and therefore "you can totally let yourself off the hook."[64] In Arendt's perspective, such feelings are "phony sentimentality in which all real issues are obscured."[65] In offering such cursory dismissals, however, neither philosopher grapples with the fact that feelings of collective guilt do not vanish once declared phony or unnecessary. Nor does either engage with *why* an emotional category that doesn't make sense within our contemporary frameworks of guilt remains so rhetorically pervasive and conceptually durable.

Perhaps in seeing a lack of alternatives to collective guilt, both Arendt and Morton propose a framework of responsibility. They do so as a matter of rhetorical exigency, having recognized that however untenable collective guilt may be conceptually, it nevertheless has persuasive implications that affect the mobilization of organized political action. Pivoting to responsibility is not only an issue of philosophical propriety, it is equally a matter of improved rhetorical framing. For example, in an interview with *Orion*, Morton says plainly, "Guilt is about individuals. Global warming is a billions-of-people scale problem. Let's have a conversation about being responsible instead. If you can understand something, then you're responsible for it."[66] Morton's comments echo those written by Arendt decades earlier: "What I am driving at here is a sharper dividing line between political (collective) responsibility, on one side, and moral and/or legal (personal) guilt, on the other."[67] One can have a collective responsibility, in other words, but can only be guilty on an individual basis.

When guilt is approached as an emotional ecology, the connectivity between variations of guilt, shame, and responsibility are to be expected—even embraced—rather than resisted. Acknowledging such connectivity neither negates nor undervalues the significant distinctions between them, moreover. Emotional entwinement does not erase difference, it complicates it. To propose that contemporary environmental discourse should be talking about responsibility—but not guilt—is to overlook how these categories exist in an emotional ecotone. They are reticulated phenomenologically, with rhetorics of responsibility readily conjuring feelings of collective guilt. And they are also conceptually linked, as Ashenden finds that once "we examine Arendt's arguments more closely [the] division between responsibility and guilt begins to break down."[68] When we take on collective responsibility but fail in some regard, guilt and/or shame are likely to be emotional byproducts. And in our social conditions, with plastic being so ubiquitous and normalized, failing at some level of ecological responsibility is inevitable.

Although Attenborough's call to take responsibility leaves open a wide range of interpretation, it is evident that the show as a whole prompted significant action on plastics. The BBC moved to ban all single-use plastic in their workplaces. The Scottish Parliament followed suit, announcing its decision to remove plastic drinking straws from government buildings. Popular UK restaurants have also made commitments, announcing plans to phase out plastic straws. And the Queen is acting, too, declaring an end to all single-use plastics on royal estates.

These results are remarkable, not only in their swiftness of response to *Blue Planet II*, but for how they contrast with the message being promoted by the show since its initial impact. "So many people have written saying that they want to do something and there are simple things we can do," the website *Blue Planet II: The Clean-Up* says, noting that "your responses have shown that if we start doing those small steps that are easily achievable, we can before long have an effect."[69] Attenborough appears in a video with encouraging words: "The actions of any just one of us may seem to be trivial and to have no effect, but the knowledge that there are thousands—*hundreds* of thousands—of people who are doing the same thing, that really does have an effect, so please join us."[70] The video ends by prompting viewers to "Pick an action. Make a difference," by learning about the top five ways to reduce plastic use, which includes buying a reusable water bottle, buying a bag for life, buy a reusable coffee cup, buying reusable straws, and buying a takeaway container for one's food.

A focus on *individual responsibility* is accentuated in these rhetorics, which coalesce around consumption techniques—cutting back on one's plastic use, buying items that aid in using less plastic, etc.

This pathway for reparative action is likely a familiar one. In the following chapter, "Eco-friendly Scapegoats," I attend to the rhetorical evolution of this strain of environmentalism. Before doing so, however, I conclude this chapter with a brief examination of Chris Jordan's photograph from his series *Midway*, which provides a counter-ballast to the information-dump-style, and suggests a minimalist, enthymematic approach to the rhetorical challenges of environmental guilt.

Enthymematic Invitations of Collective Guilt

The photograph in Fig. 2.4 is representative of those collected in Chris Jordan's *Midway*. Each photograph presents a single albatross, each in different stages of decay and each with a mass of plastic in the cavity where its stomach used to be. The plastic contents vary with each scene, and though there are familiar objects—milk jug rings, lighters, caps to toothpaste tubes—much of it is unrecognizable, just small piles of

Fig. 2.4 Albatross, perished by plastics pollution, Midway Islands. Photograph by Chris Jordan, from *Midway: Message from the Gyre* (2009–Current). Reprinted with permission

broken-down bits, origin unknown. Each photograph suggests the same fate: plastic received under the guise of nourishment turned lethal.

Jordan's art grapples with "the trans-scalar imaginary," which he characterizes as the struggle we have in oscillating between seeing the bigger picture and its details. On one hand, Jordan says, "we all know that the collective is made up of lots and lots of individuals, and that each of us is individually contributing to the catastrophe of global warming, the destruction of our seas, or whatever."[71] When our perspective is zoomed out, we can see the cumulative effect and easily identify the problem as profoundly collective, both its creation and impact. But when we zoom into consider how we as individuals are connected to the bigger picture, "each of our contributions is infinitesimal—incomprehensibly small and abstract," he adds.[72] Hurtling between the collective and individual while considering environmental atrocities can provoke a range of reactions, taxing us emotionally and challenging the threshold limits of our conceptual categories. In particular, our understandings of responsibility, guilt, and matters of care are transformed when shuttled between scales of the one and the many. For Jordan, a preeminent point to consider for contemporary environmental communication, is the simple fact that "it's really hard to hold onto our sense of empowerment as an individual [when] we behold the enormity of the cumulative effects of our behaviors."[73] Moving between the big picture and one's individual role in it, in other words, can induce a sense of paralysis.

Jordan invites viewers to engage with these quandaries of scale central to environmental issues in ways that differ from more common approaches. The first case analysis of this chapter illustrated how the use of numbers has been a primary means for relaying ecological crises. From this emerges the rhetorical mode of data dumping, a style that attempts to communicate severity and urgency by lashing audiences with overwhelming series of quantitative measures or comparisons. However, as was shown earlier, our cognitive capacity for comprehending large numbers is remarkably small. "Even people who are fantastic mathematicians," Scott Slovic notes, "cannot understand such [large] numbers on a visceral, emotional level."[74] Jordan's art aims at accomplishing what data dumping cannot: "I want to comprehend numbers that specifically relate to these catastrophic-level, unconscious behaviors that we collectively engage in as a culture."[75]

The photographs from *Midway* are minimalist in their presentation. To make sense of what is being presented, viewers are prompted to first link the bird's carcass with the pile of plastics, which the photo's composition makes viscerally obvious. In moving to establish the next link, viewers are drawn into an open space of contemplation. Jordan's *Midway* photographs function as a form of *visual enthymeme*, wherein two premises are provided: the decaying remains of an albatross and a pile of plastics exposed from its stomach. The two premises create an enthymematic juncture that invites viewers to complete the sequential chain of argument. Although the enthymeme is often defined as an incomplete syllogism, missing one component of its classical three-part structure, Aristotle's definition is in fact more dynamic than the fill-in-the-blank translation that populates most rhetoric textbooks. "[The] enthymeme," he writes in the *Rhetoric*, "must consist of a few propositions, fewer often than those which make up the normal syllogism."[76] The number of propositions isn't what makes the enthymeme "the substance of rhetorical persuasion," but their concision. Moreover, the premises need not be demonstrable fact, since in the Aristotelian enthymeme, as Ed Corbett explains, premises were often "merely probable [and thus] leading to conclusions that were only generally or usually true."[77]

In arguing that Jordan's photographs function as visual enthymemes, then, I am not suggesting that viewers are locked into a procedural logic, confined to filling in a single concluding component impervious to qualification. The invitational form of the enthymeme is far more dynamic, not only because of the flexibility in premises, but because its appeals engage "ethical and emotional dimensions of argument as well as the logical."[78] The affective dimensions of the enthymeme are evident in its etymological core—*thymos*—which connotes bodily and spiritual passion. Jeff Walker notes that the verb *enthymeomai* could be translated variously as "*lay to heart*, *consider well*, *reflect on*, *think deeply about*, [or] *be hurt or angry at*."[79] Walker concludes that engaging with an enthymeme might be described as "inference-making of the heart."[80] Visual enthymemes may further heighten the affective impact. In "The Rhetoric of Visual Arguments," Anthony Blair asserts that "one can communicate with much more force and immediacy than verbal communication."[81] The premises posed by Jordan's photographs—a dead bird and a pile of plastic—arrive *instantly* and *simultaneously*, before the audience has time to shore up defenses, a scenario that is much more difficult to achieve in verbal argument. The *Midway* photos are confrontational, not in an adversarial sense, but as a result of the image's immediacy combined with its challenging premises.

In *Visualizing Posthuman Conservation in the Age of the Anthropocene*, rhetorician Amy Propen similarly argues that Jordan's photographs "provide an invitation to engage, not to walk away—they are an invitation to 'stay with the trouble,' via Haraway, and in doing so, to configure new opportunities for world-making."[82] In addition to being an invitation, Propen contends the *Midway* series constitutes a "rhetorical response to the slow violence [of the Anthropocene] that enacts the productive ethical obligations that agential realism can entail."[83] She argues that the photos "provide a cautionary tale" that "seemingly take the viewer to task," which "runs the risk of sliding into moralism, though [they] do not stay there—[they] are always in flux, always full of interpretative possibility."[84] The photos serve "as a mirror of our collective, toxic consumerism" and "reveal the hegemony of human exceptionalism."[85] They perform "a paradoxical mode of 'ethical relation' that dissolves the distinction between the living and the dead, and in doing so performs a becoming *with* by way of grieving *with*" thereby helping to "enact an ethical obligation in the material becoming of the world [including] the work of mourning."[86]

I catalogue the range and capaciousness of Propen's claims on the photographs' rhetorical dynamics because they reflect how Jordan's photos, as visual enthymemes, invite viewers into complex negotiations. While I find many of Propen's claims compelling and provocative, here I am specifically drawn to the diversity of arguments made about what the photos *do*. In other words, I call attention not to the specific content of these claims per se, but rather, to their variegation. As visual enthymemes, the *Midway* photographs do minimal work, providing just two premises, creating an invitational juncture for the viewer to bring these premises together with additional components to form an argument. It is the "missing" component of the enthymeme that creates the conditions for "inference-making of the heart" and makes it such a potent rhetorical device. In arguing that the photos dissolve distinctions between life and death, enact ethical obligations, seemingly chide viewers, risk moralism, reveal hegemony, remain always in flux, etc., Propen illustrates their enthymematic power.

As Stephanie Young argues, the "visual enthymeme can help scholars to better understand the participatory nature of the audience in constructing proofs and cultural arguments."[87] Propen demonstrates how the *Midway* series acts as a potent catalyst for thinking through tangled webs of high-stakes issues. Whereas she proposes that "[the] photos seek to enact a more compositional, recuperative ethics" and "ethical obligations," I suggest instead that these are the results of enthymematic engagement. Far from nullifying any of her claims, my argument only slightly shifts the agential framework. Rather than claiming the *photos* "provide a cautionary tale" that "seemingly take the viewer to task," or that the series "runs the risk of sliding into moralism," I argue that the photos excel at generating a space that pulls viewers into an enthymematic arrangement, in which they generate a rich set of possible conclusions that insist upon completion of the set-up. In other words, it's not that the photos "are always in flux," but that viewer is obliged to experience a state of fluctuating reactions to the complexities of collective guilt.

Reviewing Communicative Complexities

As we've seen in the several examples in this chapter, guilt and its subspecies, collective guilt, are complex—moreso than prevailing emotional literacies and vocabularies can account for. The problem of scale is doubled when acts of environmental harm become attributed to humans qua humanity. In this case, feelings of collective guilt for participating in ecological harm are liable to be conflated with shame, which internalizes our focus rather than orienting us toward productive indignation and collective action.

One means of dealing with the communicative complexities of environmental guilt is evidenced by Chris Jordan's photographs that document the bodies of albatrosses killed by plastics pollution. The visual enthymemes he presents engage this complexity directly, not only by educating us about the consequences of our participation in the disposable material world economy, but also by engaging us emotionally, inviting us to bear witness to our feelings of guilt in a way that expands our sense of ecological connection and illuminates the possibilities of care. As we'll see in the next chapter, these complex feelings are often exploited to obscure and deflect issues of actual culpability.

Notes

1. Joshua DiCaglio, Kathryn M. Barlow, and Joseph S. Johnson, "Rhetorical Recommendations Built on Ecological Experience: A Reassessment of the Challenge of Environmental Communication," *Environmental Communication* 12, no. 4 (2018): 438.

2. Ibid., 440.
3. Ibid., 443.
4. Ibid., 990.
5. Joshua Trey Barnett, "The Ecological Awareness of an Anthropocene Philosopher," *Environmental Communication* 12, no. 7 (2018): 990.
6. Samantha Ashenden, "The Persistence of Collective Guilt," *Economy and Society* 43, no. 1 (2014): 55.
7. Ibid., 56.
8. Lauren Berlant, *The Queen of America Goes to Washington City: Essays on Sex and Citizenship* (Durham: Duke University Press, 1997): x.
9. George Lakoff, "Why It Matters How We Frame the Environment," *Environmental Communication* 4, no. 1 (2010): 72.
10. My definition draws on Nyla Branscombe's, who writes, "Collective guilt reflects the remorse that is felt when one's group has illegitimately harmed another group and not repaired the damage done." Nyla R. Branscombe and Bertjan Doosje, "International Perspectives on the Experience of Collective Guilt," *Collective Guilt: International Perspectives*, eds. Nyla R. Branscombe and Bertjan Doosje (Cambridge: Cambridge University Press, 2004): 3.
11. Mark Ferguson and Nyla R. Branscombe, "The Social Psychology of Collective Guilt," *Collective Emotions: Perspectives from Psychology, Philosophy, and Sociology*, eds. Christian von Scheve and Mikko Salmela (Oxford: Oxford University Press, 2014): 252.
12. Branscombe and Doosje, "International Perspectives," 4.
13. Ashenden, "Persistence," 55, 56.
14. Robyn K. Mallett, "Eco-guilt Motivates Eco-friendly Behavior," *Ecopsychology* 4, no. 3 (2012): 223.
15. Telegraph Reporters, "World's Deepest Plastic Bag Found at Bottom of Mariana Trench—Highlighting Spread of Ocean Pollution," *The Telegraph*, May 9, 2018, https://www.telegraph.co.uk/news/2018/05/09/worlds-deepest-plastic-bag-found-bottom-mariana-trench-highlighting.
16. Jane Bennett, *Vibrant Matter: A Political Ecology of Things* (Durham: Duke University Press, 2010): 21.
17. Chris Wilcox, Erik Van Sebille, and Britta Denise Hardesty, "Threat of Plastic Pollution to Seabirds is Global, Pervasive, and Increasing," *PNAS* 38 (2015): 112–118.
18. Michelle Paleczny, Edd Hammill, Vasiliki Karpouzi, and Daniel Pauly, "Population Trend of the World's Monitored Seabirds, 1950–2010," *PLoS ONE* 10, no. 6 (2015).
19. Wilcox et al., "Threat of Plastic," 116.

20. Andrés Cózar, Fidel Echevarría, J. Ignacio González-Gordillo, Xabier Irigoien, Bárbara Úbeda, Santiago Hernández-León, Álvaro T. Palma, Sandra Navarro, Juan García-de-Lomas, Andrea Ruiz, María L. Fernández-de-Puelles, and Carlos M. Duarte, "Plastic Debris in the Open Ocean," *PNAS* 111, no. 28 (2014).
21. Julien Boucher and Damien Friot, *Primary Microplastics in the Oceans: A Global Evaluation of Sources* (Gland: IUCN, 2017).
22. Marc Bain, "If Your Clothes Aren't Already Made Out of Plastic, They Will Be," *Quartz*, June 5, 2015, https://qz.com/414223/if-your-clothes-arent-already-made-out-of-plastic-they-will-be/.
23. Chris Tyree and Dan Morrison, "Invisibles: The Plastic Inside Us," *Orb Media*, https://orbmedia.org/stories/Invisibles_plastics.
24. Janet Raloff, "How Plastic We've Become," *ScienceNews*, January 17, 2008, https://www.sciencenews.org/blog/food-thought/how-plastic-weve-become; *Fourth National Report on Human Exposure to Environmental Chemicals*, Center for Disease Control and Prevention, February 2015, https://www.cdc.gov/biomonitoring/pdf/FourthReport_UpdatedTables_Feb2015.pdf.
25. Claire Beausoleil, et al., "Regulatory Identification of BPA as an Endocrine Disruptor: Context and Methodology," *Molecular and Cellular Endocrinology* 475, no. 5 (2018).
26. Timothy Morton, *Being Ecological* (Cambridge: MIT Press, 2018): xviii.
27. Frederikke Magnussen, "A Dolphin Feeds on a Floating Bag," *Daily Mail Online*, March 8, 2018, www.dailymail.co.uk/news/article-4120622/A-dolphin-feeds-floating-bag.html/.
28. Paul Slovic and Daniel Västfjäll, "The More Who Die, the Less We Care: Psychic Numbing and Genocide," *Numbers and Nerves: Information, Emotion, and Meaning in a World of Data*, eds. Scott Slovic and Paul Slovic (Corvallis: Oregon State University Press, 2015): 27.
29. Ibid.
30. Joseph Curtain, "Let's Bag Plastic Bags," *The New York Times*, March 3, 2018, https://www.nytimes.com/2018/03/03/opinion/sunday/plastic-bags-pollution-oceans.html.
31. Scott Slovic and Paul Slovic, "Introduction: The Psychophysics of Brightness and the Value of a Life," *Numbers and Nerves: Information, Emotion, and Meaning in a World of Data*, eds. Scott Slovic and Paul Slovic (Corvallis: Oregon State University Press, 2015): 2.
32. Ibid., 4.
33. Morton, *Being Ecological*, xviii.
34. Ibid., xix.
35. Slovic and Västfjäll, "The More Who Die," 31.

36. Nick Triolo, "Four Questions for the Author: Timothy Morton, Being Ecological," *Orion Magazine*, September 22, 2018, https://orionmagazine.org/2018/09/four-questions-for-the-author-timothy-morton-being-ecological/.
37. Sam Littlefair, "Groundbreaking Scholar Timothy Morton Wants Philosophers to Face Their 'Buddhaphobia'," *Lion's Roar: Buddhist Wisdom for Our Time*, September 2, 2017, https://www.lionsroar.com/groundbreaking-scholar-timothy-morton-wants-philosophers-to-face-their-buddhaphobia/.
38. Morton, *Being Ecological*, xiii.
39. David Attenborough, "David Attenborough: We Must Act Now to Protect Our Oceans from the Deadly Threat of Plastic," *RadioTimes*, accessed March 1, 2018, https://www.radiotimes.com/news/tv/2018-03-16/david-attenborough-blue-planet-ii-we-must-act-now-to-protect-our-oceans-from-the-deadly-threat-of-plastic/.
40. Ibid.
41. *Blue Planet II*, 7, "Our Blue Planet," produced by James Honeyborne, aired December 10, 2017, on BBC One.
42. Ann Cvetkovich, *An Archive of Feelings: Trauma, Sexuality, and Lesbian Public Cultures* (Durham: Duke University Press, 2003): 10.
43. Ibid., 7.
44. Sara Ahmed, *The Cultural Politics of Emotion* (Durham: Duke University Press, 2004): 10.
45. Ben Highmore, *Cultural Feelings: Mood, Mediation and Cultural Politics* (New York: Routledge Press, 2017): 65. Emphasis added.
46. Alex Lockwood, "Graphs of Grief and Other Green Feelings: The Uses of Affect in the Study of Environmental Communication," *Environmental Communication* 10, no. 6 (2016): 738.
47. Ahmed, *Cultural Politics*, 10.
48. Deborah Tollefsen, "The Rationality of Collective Guilt," *Midwest Studies in Philosophy* 30, no. 1 (2006): 228. Emphasis added.
49. Brian Lickel, Toni Schmader, and Marchelle Barquissau, "The Evocation of Moral Emotions in Intergroup Contexts: The Distinction Between Collective Guilt and Collective Shame," *Collective Guilt: International Perspectives*, eds. Nyla R. Branscombe and Bertjan Doosje (Cambridge: Cambridge University Press, 2004): 41.
50. Ibid., 230.
51. Cf. Karen P. Leith and Roy F. Baumeister, "Empathy, Shame, Guilt, and Narratives of Interpersonal Conflicts: Guilt-prone People Are Better at Perspective Taking," *Journal of Personality* 66, no. 1 (1998): 1–37.
52. Attenborough, "We Must Act Now," *RadioTimes*.

53. I am sensitive to concerns over using the Holocaust as a comparative reference point for analyzing the mass extinction we're currently undergoing. Ultimately, however, I concur with Derrick Jensen, who argues that isolating the Holocaust as an aberration without parallels to other mass exterminations "allows meaningful analysis to stop [and such] shunting of attention is necessary if we're going to continue to live the way we do….To pretend the Holocaust is Unique is to not question our own way of living, to not question our own innocence." *The Culture of Make Believe* (White River Junction: Chelsea Green Publishing, 2004): 564.
54. Hannah Arendt, "Collective Responsibility," *Amor Mundi: Explorations in the Faith and Thought of Hannah Arendt*, ed. James W. Bernauer (Boston: M. Nijhoff, 1987): 43.
55. Ibid., 43–44.
56. Ibid., 43.
57. Hannah Arendt, *Eichmann in Jerusalem: A Report on the Banality of Evil* (New York: Viking Press, 1963): 276.
58. Arendt, "Organized Guilt," 148–149.
59. Ibid., 149.
60. Arendt, *Eichmann*, 247.
61. Ashenden, "Persistence," 62.
62. Arendt, "Collective Responsibility," 43.
63. Morton, *Being Ecological*, xiii.
64. Ibid.
65. Arendt, "Collective Responsibility," 44.
66. Triolo, "Four Questions," *Orion*.
67. Arendt, "Collective Responsibility," 47.
68. Ashenden, "Persistence," 64.
69. *Blue Planet II: The Clean-Up*, accessed May 13, 2018, https://canvas-story.bbcrewind.co.uk/sites/blue-planet-two-six-months-on/
70. Ibid.
71. Scott Slovic, "Introspection, Social Transformation, and the Trans-scalar Imaginary: *An Interview with Chris Jordan*," *Numbers and Nerves: Information, Emotion, and Meaning in a World of Data*, eds. Scott Slovic and Paul Slovic (Corvallis: Oregon State University Press, 2015): 205.
72. Ibid., 205.
73. Ibid., 205.
74. Ibid., 208.
75. Ibid., 209.
76. Aristotle, *Rhetoric*, eds. W. Rhys Roberts, Ingram Bywater, and Friedrich Solmsen (New York: Modern Library, 1954): 1357a.
77. Edward P.J. Corbett, "Introduction," *The Rhetoric and Poetics of Aristotle*, trans. W. Rhys Roberts (New York: Random House, 1984): xviii–xix.

78. Ibid., xix.
79. Jeffrey Walker, "The Body of Persuasion: A Theory of the Enthymeme," *College English* 56, no. 1 (1994): 49.
80. Ibid.
81. Anthony Blair, "The Rhetoric of Visual Arguments," *Defining Visual Rhetorics*, eds. Charles A. Hill and Marguerite Helmers (Mahwah: Erlbaum Associates, 2004): 53.
82. Amy Propen, *Visualizing Posthuman Conservation in the Age of the Anthropocene* (Columbus: Ohio State University Press, 2018): 51.
83. Ibid., 75.
84. Ibid., 57, 67–68.
85. Ibid., 54, 59.
86. Ibid., 74.
87. Stephanie L. Young, "Running Like a Man, Sitting Like a Girl: Visual Enthymeme and the Case of Caster Semenya," *Women's Studies in Communication* 38, no. 3 (2015): 345.

CHAPTER 3

Eco-friendly Scapegoats

Abstract This chapter demonstrates how collective guilt can be wielded to undermine the potential of and desire for collective action—how, paradoxically, collective guilt can be a powerful individualizing force. Through analysis of public relations campaigns and marketing, I illustrate how corporations have intentionally employed rhetorics of personal responsibility to contain political pressure and diffuse demands for more corporate accountability. The strategic dispersal of accountability into individual acts of "eco-friendly" consumerism both provoke and alleviate collective guilt for the individual, yet mitigate environmental damage in only the most minimal and marginal of ways. Collective guilt, I argue, calls for collective action.

Keywords Eco-friendly · Collective guilt · Scapegoating · Kenneth Burke · Double-bind · Hypocrite's trap

Save the planet, one napkin at a time. We can all make a difference: please recycle this bottle. Consider the environment before printing this email. Water is life: please take shorter showers. Upgrade to energy-efficient lightbulbs. People start pollution, people can stop it. Unplug appliances when not in use. Reduce your air travel. Recycling is smart: do your part. Turn the water off while brushing your teeth. Buy a

T. Jensen, *Ecologies of Guilt in Environmental Rhetorics*,
Palgrave Studies in Media and Environmental Communication,
https://doi.org/10.1007/978-3-030-05651-3_3

low-flow toilet. Adjust the thermostat. Choose to reuse. To feel fantastic, use less plastic. Small actions can make a big difference. It all adds up.

It has been over a decade since Al Gore's documentary *An Inconvenient Truth* concluded its case—global warming is a threat to all life on earth—with a list of actions that is likely to feel quite familiar, even quaint: buy energy-efficient appliances and lightbulbs; use a clock thermostat to reduce energy consumption; recycle; when you can, walk or ride a bicycle or use mass transit, etc.[1] Gore's list bears eerie resemblance to recommendations made in the "Do-It-Yourself Ecology" section of a booklet distributed on the first-ever Earth Day, 1970. Suggestions include "take your own shopping bag to the store," "ride a bicycle," and "turn off unnecessary lights." In the half-century span between the launch of Earth Day and the book you're currently reading, rhetorical variations of *reduce*, *reuse*, *recycle*, and *do your part* have remained remarkably consistent in their angle of appeal, even as the earth's population has doubled in size. What has changed is their pervasiveness. Eco-friendly rhetorics have thoroughly saturated everyday discourse, and can now be found everywhere from napkins to government websites.

"Eco-friendly" rhetorics—appeals loosely grouped around making small adjustments to one's everyday lifestyle for the betterment of our planet's health—constitute one of the most prominent strains of contemporary environmental communication. It's not considered strange in the least for a plastic yogurt cup to implore an act of epic heroism, calling on you to *Save Our Planet!* through the mere act of placing the cup in a recycling bin. Appeals to "green" one's daily habits and consumer choices are so constant and expected, however, that analysis of individual examples risks missing how the rhetorical force of eco-friendly rhetorics abides in their accumulative effect. Their capacity to convince, in the words of Kenneth Burke, is owed "much more to trivial repetition and dull daily reinforcement than to exceptional rhetorical skill."[2]

Although eco-friendly rhetorics are not confined to consumer goods and services, the bulk occur in advertising and institutional branding, where their presence is ubiquitous. As awareness of environmental issues has grown over time, so has the use of green marketing, a pattern that shows no signs of abatement. A 2015 Nielson poll of 30,000 people across 60 countries found that 63% of consumers are willing to pay more for eco-friendly brands; for Millennials, that figure jumps to 73%, a 50% increase from the year prior.[3] Another recent marketing report indicates "both U.S. consumer awareness and consumer attitudes toward sustainable, or 'green,' brands have increased to an all-time high."[4] Being environmentally friendly is big business—simultaneously a consequence of and cause for the prevalence of eco-friendly rhetorics.

That "green" and "environmentally friendly" tropes are used to leverage public interest in environmental issues into selling products and services is palpably self-evident. Likewise, it is an axiom of this chapter that eco-friendly rhetorics frame consumption management as the preeminent pathway for engaging and advancing social causes. They help reify neoliberal logics that "imbue the market with ethical potential and social responsibility and the public realm with market metrics," as Wendy Brown remarks.[5] My aim is to advance our understanding of *how* this conspicuous strain of environmental communication entrenches the neoliberal precepts of individual responsibility and market-based reasoning. My argument is that eco-friendly rhetorics steadily stoke low-lying levels of collective guilt for one's complicity in an economic system that is ecologically destructive, and then, in perverse irony, invite one to atone for their guilt via market-based solutions and consumption management techniques. Eco-friendly rhetorics perpetuate, in short, a pernicious guilt-atonement-cycle. Moreover, they function in devilishly symbiotic fashion, such that the recurring loop of collective guilt and consumer atonement helps to further circulate eco-friendly rhetorics.

Kenneth Burke's scapegoat schema provides a useful lens for understanding this cycle, though only up to a point. Burke's model underscores how scapegoat rhetorics rely on drawing hard boundaries between "they" and "us." In contrast, I illustrate how an eco-friendly scapegoat strategy obscures accountable agents by *expanding* the "we." The plastic bottle label that reminds you *We can all make a difference* and therefore *Please recycle*, functions as a pin-prick of collective guilt—infinitesimally small, likely to go unnoticed—followed by an equally faint offer of easy atonement. As a singular instance, this cycle has a trifling impact. In the aggregate, eco-friendly rhetorics bring about a patterning of public feeling that inhibits effective action on ecological issues. Friendliness, I lament to inform you, is not the affective incentive of eco-friendly rhetorics.

This chapter therefore aims to sensitize us to the rhetorical transfer of blame within an age of ecological upheaval. I trace the germinal seeds of this scapegoat strategy back to the emergence of the environmental movement to illustrate its enduring effectiveness. From the Crying

Indian PSA in 1971 to the BP Deepwater/Horizon spill in 2010 to critiques of the Green New Deal in 2019, a set of rhetorical maneuvers has consistently been used to safely contain environmental action and redirect emotional energy back into the market. In each instance, scrutiny of corporate accountability is deflected back onto the individual consumer, a transfer of blame that I argue is assisted by the pervasive circulation of eco-friendly rhetorics. Although eco-friendly messages ostensibly conjure the collective through their use of a vague "we," these rhetorics postulate an independent and self-determining individual governed by rational faculties. This figuration is essential to eco-friendly scapegoating and the neoliberal logics they operate in conjunction with, for it allows complicity in systematic environmental destruction to be framed as *voluntary*—the result of intentional, reasoned choices. And the individual in this configuration is one that can be held culpable, as they *consciously choose* to participate in ecological harm.

My progression through these ideas takes a chronologically circular format. I begin not in 1971, with the emergence of a nascent environmental movement and the corporate strategies to contain its impact, but in 2019, with the fruits those strategies have propagated. Building off the work of scholars who have skillfully mapped prevalent pitfalls and rhetorical snares that environmental advocates routinely face today, I analyze the emotional dynamics that make those traps cohesive and effective. I first examine the "hypocrite's trap," which brandishes complicity to silence critique, and then consider its close kin, the double-bind. To situate these rhetorical binds within an ecology of environmental guilt, I analyze the Crying Indian PSA, whose debut occurs at a kairotic moment in the struggle to establish the orthodoxy of environmentalism. By instrumentalizing collective guilt and diverting attention from corporate accountability, the PSA lays a foundation for the ascendance of eco-friendly rhetorics. Coming full circle, I show how corporations continue to employ containment strategies anchored in collective guilt. Along with tactics like the hypocrite's trap, these techniques are constellated within larger ecologies of environmental guilt, which are heavily influenced by the ubiquity of eco-friendly messages. As a principal means of reinforcing an individualistic orientation toward addressing ecological crisis, *eco-friendly rhetorics maintain conditions that ultimately undermine calls for meaningful change*—regardless of whether proposals call for individual or collective action.

What is at stake here for the study of environmental communication is the recognition of how collective guilt can be wielded to undermine the potential of and desire for collective action. Paradoxically, collective guilt can manifest as a powerful *individualizing force*. What I find particularly insidious about eco-friendly scapegoating, however, is its capacity to obscure opportunities for thinking ecologically, thereby attenuating the potential for practicing ecological care. Ecological attunement expands the individual, rendering one radically interdependent, a constitution of relations. Illuminating how environmental guilt is actively constructed, then, is one step toward a larger goal: to open a space for reorienting our relationship to blame and responsibility through ecological attunement and an ethic of reciprocal care.

The Hypocrite's Trap and Other Inevitabilities

On February 7, 2019, Representative Alexandria Ocasio-Cortez (AOC) and Senator Ed Markey submitted to the House of Representatives Resolution 109, "Recognizing the duty of the Federal Government to create a Green New Deal."[6] The resolution proposes a package of bold reforms to American infrastructure, travel, and agriculture through a "national, social, industrial, and economic mobilization on a scale not seen since World War II and the New Deal era."[7] Over a span of ten years, it aims to achieve, among other items, "net-zero greenhouse gas emissions … clean air and water; climate and community resiliency [and] a sustainable environment."[8] Intended primarily as a statement of priorities and legislative inspiration, the "Green New Deal" (GND) quickly became a flash point of political and environmental communication. Although the idea of a Green New Deal had been floating about in various arenas for roughly a decade, the attention paid to this version was greater in both quantity and intensity. The combination of a deeply partisan political setting, a lofty set of goals on an issue that divides political parties, and the fierce energy of a newly elected Ocasio-Cortez ignited substantial commentary infused with a mix of enthusiasm and skepticism, delight and disdain.

Among the recurring tropes and frames deployed in reactions to the GND is the *hypocrite's trap*—a rhetorical construction emphasizing how the proposer participates in and/or benefits from the system they critique and seek to change. Consider this headline from a *Fox News* article published two days after the GND's release: "2020 Democrats jump to endorse Green New Deal despite spending hundreds of thousands on air travel—including private jets."[9] The article mentions Senator Kamala Harris's support for the GND, citing her statement, "We must aggressively tackle climate change which poses an existential threat to our nation," then immediately notes, "Yet Harris herself is far from following what she preaches. Since 2015, her campaign has spent around $300,000 on air travel."[10] The same maneuvers are repeated by the *New York Post*, who ran a cover with the headline, "Gas-guzzling car rides expose AOC's green hypocrisy," followed by "ECO TRIP" in all caps.[11] "The Democratic firebrand," it decries, "took Amtrak far less [than her use of cars and airplanes] despite high-speed rail being the cornerstone of her save-the-world strategy."[12] The credibility of both proposal and proposer are evaluated by how closely personal actions align with recommendations for altering social systems. By emphasizing gaps between the two, charges of hypocrisy—faking the role of a virtuous person, while failing to live by the principles they chasten others to uphold—can be leveled, along with the recommendation to lead by example.

The hypocrite's trap is adroitly analyzed by rhetoricians Jen Schneider, Steve Schwarze, Peter Bsumek, and Jen Peeples in their book, *Under Pressure: Coal Industry Rhetoric and Neoliberalism*. There, they define the hypocrite's trap as "a set of interrelated arguments that attempts to disarm critics of industries that provide particular goods or technologies based on the critics' own consumption of or reliance on those goods."[13] In tracing how the hypocrite's trap is used by industry groups and other professionals in attempting to erode the validity of fossil fuel divestment efforts, Schneider et al. breakdown the trap's construction into three steps: establish ignorance; expose complicity; and shame hypocrisy.[14] In each of the three steps, they illustrate, the divestment movement is characterized as "idealistic and unrealistic, elitist and dangerous, and hypocritical and amoral."[15] The same portrayals can be found in the GND examples: the reference to "private jets" signals elitism; the phrase "her save-the-world strategy" suggests naïve idealism; and the invocation to "practice what you preach" exclaims the disingenuousness and hypocrisy of GND supporters.

The rhetorical impact of the hypocrite's trap is registered not only in these surface-level disparagements, but also, Schneider et al. note, through the ideological work it performs in "naturalizing and normalizing neoliberalism as common sense."[16] David Harvey concurs that the persuasive power of neoliberalism exists at the level of tacit, everyday logics. He begins *A Brief History of Neoliberalism* with the contention that this instantiation of capitalism has "pervasive effects on ways of thought to the point where it has become incorporated into the common-sense way many of us interpret, live in, and understand the world."[17] Jason Read similarly argues, "[it] is not the structure of the economy that is extended across society, but the subject of economic thinking, its implicit anthropology."[18] Neoliberalism reproduces itself by promoting marketplace rationalities as the most credible, if not the only valid means of logic for twenty-first-century living.[19] As a vehicle for containing activist pressure and reinscribing neoliberal norms, "the hypocrite's trap is designed to economize politics by individualizing responsibility," Schneider et al. explain.[20] It enacts this ideological work through two interrelated channels, first by "exposing complicity" and second, by "shaming hypocrisy," the combination of which works to achieve the "hallmarks of neoliberalism: folding the political into the economic, the collective into the individual."[21]

Results are often effective. The authors concede that the coal industry's rhetorical strategies are "both wily and agile," often able to topple environmental advocates off the moral high ground, and then claim that mantle for themselves. Schneider, Schwarze, Bsumek, and Peeples provide deep—and deeply useful—analytical insight into the tried-and-true tactic of the hypocrite's trap. Understanding why this strategy is so effective, however, requires additional explanation. Industry advocates and skeptics of environmental advocacy may cast charges of complicity and hypocrisy, but *why* those indictments are so compelling—why they disrupt activist momentum, seeding doubt and opprobrium of environmental agendas and advocates—necessitates further consideration of the rhetorical and emotional ecologies in which those accusations circulate.

In expanding the rhetorical ecologies of complicity, it is valuable to consider how the hypocrite's trap is constructed by those other than industry lobbyists. Schneider et al. perform some of this work by including examples from university presidents and officials. They describe a meeting at the University of Colorado, Boulder, for example, in which students called on the Board of Regents to divest the University's endowment from fossil fuel companies. One Regent asked if students drove to the meeting, implying that their consumption of gas in route to a meeting advocating we collectively consume less gas undermined the authenticity of their argument. "[How much gas] could've been saved by a presentation up there on that screen?" the Regent asked, adding, "I'm asking you to use those solutions yourselves. It gives more credibility to these statements."[22] Although they may not deploy it with the same motivations of pro-coal boosters, university officials in this example nevertheless engage the hypocrite trap's root logic and imitate key aspects. These examples suggest that the hypocrite's trap is realized along various spectra, such that gradations of intentionality, awareness, and adaptation will affect how the trap is constructed and received.

Accusations of environmental complicity and hypocrisy extend well beyond those lobbed by corporate interests, university executives, and political pundits, of course. Environmental activists often apply the trap against other environmentalists. We often enact it against ourselves. Those of us engaged with ecological issues routinely invoke the three steps laid out by Schneider et al., criticizing our own ignorance, complicity, and hypocrisy. Cultivating an ecological perspective often propels us toward these steps, though in a very different vector and through other emotional valences. For many of us, our knowledge of complicity is more extensive than any coal lobbyist or university regent could offer. The regent who proposed a virtual presentation as a "greener" alternative fails to see how using the electrical grid also invites complicity with ecological destruction. Even in Boulder, Colorado, a town committed to reaching 100% renewable energy by 2030, the electrical grid is currently fueled in part by fracked gas.[23] Delivering a virtual presentation links one to hydraulic fracking and its ecological legacy of air pollution, water contamination, earthquakes, and radioactive exposure. Ecological awareness is a prerequisite for recognizing complicity and as it grows, the webs of complicity are spun faster and thicker.

Schneider et al. rightly note that "the complicity argument shuts down the debate because almost none of us can claim to be free of dependence on fossil fuels."[24] Complicity arguments, however, also open up debate and inquiry; where they lead is the most pressing issue. In the following section, I look at a double-bind of environmental advocacy that is inextricably connected to the hypocrite's trap. If participating in a system that causes environmental damage renders one complicit, it follows that to critique systematic harm, one must "lead by example" and "practice what they preach" by extricating themselves from it. The "No Impact Man" project (NIM), a year-long experiment to achieve zero environmental impact, is an attempt to do just that. The project takes the directive of eco-friendly rhetorics—make adjustments to your everyday lifestyle for the betterment of our environment—and pushes its logic to various thresholds. My analysis of No Impact Man serves as a link between the hypocrite's trap and this chapter's larger argument, that eco-friendly rhetorics maintain conditions that ultimately undermine calls for meaningful change.

On the Limits of No Impact

"The average American family creates 1,600 lb of trash a year. We're trying to reduce that to zero," states Colin Beavan in the film, *No Impact Man*. In attempting to have a net-zero environmental impact for one year while living in New York City, Beavan and his family went to extremes to reduce not only their trash to zero, but also to achieve "zero carbon … zero pollution in the air, zero resources sucked from the earth, zero toxins in the water."[25] In addition to the documentary, the experiment's trials and insights were recorded through Beavan's blog as well as a book. The protracted title of the latter gestures to how project's impetus is framed: *No Impact Man: The Adventures of a Guilty Liberal Who Attempts to Save the Planet and The Discoveries He Makes About Himself and Our Way of Life in the Process.*

Beavan unfolds the origin story of the NIM project in the book's introduction, noting, "[we] feel a malaise and a guilt that at another time in history might have motivated action, but that this time seems instead to be coupled with a terrible sense of helplessness."[26] This enfeebling environmental guilt, he later suggests, develops through slow accrual:

> The truth was that every coffee cup and every water bottle in the corner trashcan gave me a tiny micro-twinge of guilt. For years, I'd given myself a daily pass with the intention of eventually doing better. If you had asked me if I tried not to make trash, not to waste, I would have told you that I certainly didn't produce the average American's 4.6 lb of trash per day, or roughly 1,700 lb per year. I would have probably told you I didn't try as hard as I should but that I tried. I made an effort, I would have said. I'm not all talk. I care about the world.[27]

The liminal state Beavan describes captures the subtle, everyday environmental guilt this book seeks to spotlight. The slow piling of *almost* negligible guiltings that register barely and rarely. The positive reinforcement of self-identity. The blend of confession and defensiveness in response to the suggestion of hypocrisy. The easy-to-rationalize waste. The environmental guilt he describes is a diminutive but persistent guilt, whose management is well-practiced and often swept away swiftly thanks to a wide range of coping techniques. It occurs in "modes of lower case drama," to adopt the phrasing of Lauren Berlant.[28] In theorizing the widely shared feelings that pattern social life, Berlant emphasizes their minor registers, the "pulsations of habituated patterning that make possible getting through the day."[29] The "micro-twinges" of environmental guilt are common, shared by many, yet often difficult to address because of their subtlety. As Ben Highmore asks, "How, for instance would you *directly* respond to the low, long drone of anxiety of nuclear war?"[30] NIM suggests that the small pin-pricks and micro-twinges can accumulate and lead to larger ruptures.

But there are other readings of the origin of NIM. In her review, Elizabeth Kolbert, respected environmental science writer and author of the Pulizter-winning *The Sixth Extinction: An Unnatural History*, writes with incredulity, "[t]here's something a tad disingenuous here. Beavan is, after all, a man whose environmental activism began over lunch with his agent."[31] Authenticity of motivations is one of several evaluations Kolbert makes in auditing NIM. She calls foul on Beavan's "experimental rigor," pointing out that his blog stays running, which uses electricity. She alludes to the economic privilege underwriting the project and expresses exasperation at Beavan's disgust of items individually wrapped plastic when millions of people live in poverty. What the project is gravely missing, Kolbert argues, is perspective—the ability to "see the deforestation for the trees."[32] The focus on individual action, she contends, distracts from the measures of collective action needed to solve complex, global problems.

These critiques are echoed across the majority of reviews of NIM. NPR's review finds flaws in the project's thoroughness, noting that "Beavan won't surrender everything modern, which undercuts his experiment."[33] *The New York Times* opens their review of Beavan's book by noting its material impact: "There's a certain problem with branding oneself a radical environmentalist superhero and then letting a real old-fashioned book about the experience roll luxuriously off the presses."[34] No Impact Man, in short, is critiqued for making an impact (materially, through the book's printing or his laptop's electricity) and for *not* making an impact (as a result of focusing on individual actions, rather than collective solutions).

In "The Impact of 'No Impact Man': Alternative Hedonism as Environmental Appeal," Jen Schneider and Glen Miller call attention to NIM's double-bind. "NIM needs to obtain media coverage, which may require being 'radical' or 'extreme,' and at the same time promote sympathy for environmental activism without losing the mainstream audience's attention," they write, detailing one of several ways in which NIM is caught in a double-bind.[35] The authors source their use of term "double-bind" to an article on climate change communication in popular media outlets, where scientists are often forced to choose between "loyalty to the scientific method," which calls for caveats, qualifiers, and technical language, and successfully delivering their message through channels that "demand succinct explanation, familiar images, and metaphors."[36] Similar to NIM, climate change scientists must choose between speaking in a context that does not register their message or not engaging at all, another version of failing to communicate.

The origin of the double-bind as theoretical concept is traced to Gregory Bateson, a polymath who developed it as part of his psychiatric research in the 1960s. For Bateson, a double-bind is a communicative paradox in which a directive is refuted by its context: "The *context* was an imperative; the message ordered the listener to ignore that context. One could neither comply nor escape complying."[37] Different from a no-win situation, in which one is asked to choose between two poor options, the double-bind works to erase any framework of choice. Bateson immediately saw the concept's relevance for addressing environmental issues. It captures "the overwhelming trapped-ness and runaway still associated with ecological crisis—that painful experience when the 'correct' response to environmental problems seems to produce no solution but only to exacerbate those problems at an accelerating rate."[38] The double-bind accurately describes the rhetorical ecology constructed by eco-friendly rhetorics, which urge one to use fewer napkins and periodically ride a bike instead of drive. The realities of escalating extinction rates and the breakdown of our climate, however, form a context that discounts the eco-friendly imperative. A self-negating loop is formed: it's crucial that you act—the act doesn't make an impact.

Beavan takes the directive of eco-friendly rhetorics to reduce, reuse, and recycle to their comical extreme, yet readily concedes that while his family derived many benefits from the experiment, the individual acts in no way lessened the collective problem. This is a hard-won realization that could have prompted conversation on the environmental double-bind. Instead, reviews routinely critiqued NIM for its hypocrisy on several levels, censuring the project for being not being extreme enough and for being too extreme, threatening to alienate mainstream audiences. These critiques illustrate the limits of NIM's approach to communicating environmental crisis. The authors of "The Hypocrite's Trap" delineate this dilemma:

> Activists who attempt to strictly limit their fossil fuel consumption, live off the grid, grow their own food, only wear natural fibers, or no own cars are characterized as too extreme, too far removed from 'normal' American life. Thus, their voices are also dismissed from decision making. Furthermore, in a carbon-intensive economy it is hard to imagine how an activist would disseminate their message or organize to fight fossil fuels without also consuming them. Questioning fossil fuels then makes one either a hypocrite or a radical outsider.[39]

Or both, as the critiques of Beavan illustrate. Here, the difference between a lose-lose and a double-bind is clearer. The former suggests that one *chooses* to be either a hypocrite *or* a radical outsider. In the environmental double-bind, hypocrite is default status. From there, depending on how one engages ecological problems, one exposes themselves to critiques that further delimit rhetorical capacity. The hypocrite's trap, then, must be either extended to accede that we're always-already in it (hypocrites trapped?) or narrowed to reference only explicit and targeted invocations of environmental hypocrisy.

The most repeated critique of NIM calls out the individualist approach to solving collective environmental issues, despite Beavan's evolution over the course of the project toward more collective forms of action. Near the conclusion of the documentary, he explicitly articulates the problem within a collective frame and implies the solution lies there, too: "I believe very strongly that a lot of the environmental problems in our planet have come because of the breakdown of community; because without community, none of us feels accountable to anybody else."[40] Marilyn DeLaure attributes this discrepancy to NIM's adoption of comedic frame, which conflicts with the melodrama of tragedy that viewers have come to associate with environmental narratives. Writing in *Environmental Communication*, Delaure argues that comedy "presents problems as arising from human limitations and mistakes, rather than from inherent evil," whereas tragedy "locates blame with an Evil Other [and therefore] precludes audiences from understanding themselves as agents responsible for both problem and solution."[41] Consequently, critics miss the "both/and" message of NIM, which calls for *both* individual *and* collective action.

Schneider and Miller skillfully read NIM's implicit unmasking of the double-bind, noting that the project "makes visible the hidden ways in which our behaviors, which frequently harm the environment, are often already prescribed or dictated" because our "existing infrastructure of consumption [has] embedded unsustainability into the American way of life."[42] Why, then, didn't reviewers and audiences pick up on this exposé? What rhetorical forces veil NIM's unveiling? In addition to the factors already offered by DeLaure and Schneider and Miller, I argue that eco-friendly rhetorics play a critical role in constructing a double-bind for environmental action. Their pervasive circulation in everyday consumer discourse helps conceal the infrastructure of consumption through a process I detail in the sections that follow.

In taking the promises of an eco-friendly lifestyle and amplifying them to eleven, NIM chronicles myriad examples of systematic bias toward environmental harm. Yet critiques of NIM reveal more about the traps and binds of environmental advocacy in neoliberalism than they do about the actual project. Namely, we witness how effortless charges of hypocrisy and failed purity are in systems that default everyone into a position of complicity. Industry lobbyists are just one demographic that employ the hypocrite's trap to discredit environmental advocacy. Reviewers, consumers, and environmentalists use it too. The more pervasive and persuasive enactments of the hypocrite's trap, I argue, are those we employ on ourselves in subtle and mundane ways. However, as I show in the remainder of this chapter, it's not entirely accurate to call it "self-trapping." A rhetorical ecology of eco-friendly messages activate a transfer of blame that seems so unremarkable it often escapes notice.

Containment and Emerging Environmentalism

The pair of sentences is small and so nondescript it's easy to miss. Tucked beside the nutritional information on this plastic bottle of water, it reads: "We can all make a difference. Please recycle." This particular bottle is sold by Ice Mountain, a brand of Nestlé. The water inside is pumped from the same wells that citizens of Mecosta County, Michigan use. On the front label it reads: "100% Natural Spring Water."

In 2016, bottled water surpassed soda for the most consumed beverage in the United States.[43] The same year saw 480 billion plastic drinking bottles sold globally, a number that is impossible to envision, yet one that nevertheless has profound ecological impact.[44] The yearly output is projected to reach just shy of 600 billion bottles by 2021, with bottled water sales as a driving contributor.[45] Currently, 91% of plastic bottles go unrecycled.[46] Put differently, 91% of the more than 500 billion bottles sold this year will end up in a landfill, swept out to sea, or otherwise discarded across the land.

There is a connection between the insipid, back-of-the-label statement proposing solidarity in and inspiration for recycling and the billions of bottles that are produced and then go unrecycled each year. To understand why a small statement like, "we can all make a difference—please recycle," plays a big role in contemporary environmental rhetorics, one that makes a material difference on biophysical ecologies, we can turn to the origins of the environmental movement, well before the phrase "eco-friendly" enters everyday circulation. The germinal seeds of eco-friendly acts and attitudes can be unearthed through analysis of the iconic "Crying Indian" Public Service Announcement, wherein a single tear seared an environmental ethos into the collective imagination, the influence of which continues to this day.

"In the 1950s, 1960s, and 1970s, corporate managers and government bureaucrats often tried to crush ecological protests," writes Timothy Luke in *Ecocritique*.[47] In the wake of World War II, environmental issues were not a priority in politics and barely registered in public discourse. In a booming postwar economy, corporations felt little need to be environmentally friendly—neither in public relations, nor in actual practice. Following the publication of Rachel Carson's *Silent Spring* in 1962, concerns about pesticides (which Carson suggested should be called *biocides*) and their effect on human and ecosystem health, a more robust environmental consciousness began to grow.[48] Corporations mostly viewed public outcry over unchecked environmental destruction as a nuisance to be quelled. "[F]iguring that such direct strategies of political intervention rhetorically might roll back the symbolic assaults of troublesome 'tree-hugging' nature lovers," corporations attempted to steamroll protest with the same self-promoting arguments they relied on for decades: "by continuing to convince the general public that belching smokestacks still were signs of material progress and not the stigmata of industrial pollution."[49]

As the 1960s came to a close, however, the environmental movement, though still fledgling, is a recognizable rhetorical entity—that is, people are referring to its collected acts and attitudes as a *movement*.[50] A 1970 Sierra Club Bulletin claims that, "The Environmental movement is coming to be more than a re-labeled Conservation Movement."[51] An activist 'zine, *Focus Midwest*, writes, "[t]here has been considerable concern in the peace movement about the booming Environmental movement. Some feel that this 'new' cause is siphoning off energy from the civil rights and peace movements."[52] Riding this tide of rising environmental concern, "Earth Day" was launched in 1970 to highlight environmental issues and the dangers of ecological degradation and toxification. Modeled on the "teach-ins" so prevalent in anti-Vietnam War demonstrations, the first Earth Day saw millions of participants across the United States, mostly on high school and college campuses. For many, this enormous display of organization and support marked the moment when the Environmental movement emerged as a fully legitimate social movement. The platform and agenda of that movement, however, were heterogeneous and hotly debated.

Around the same period a recognition emerged from corporate public relations circles that this social movement would prove a formidable force, worthy of a tougher response. "To contain the emergence of new, and perhaps more radical, measures of ecological transformation," Luke writes, "the corporate-run circuits of mass consumption refunctioned the rhetoric of that faction in the ecology movement."[53] This shift in corporate communication strategy can be seen in the Crying Indian PSA, which captured significant attention at a critical juncture in the Environmental movement's evolution and whose iconic image of a single tear rolling down Iron Eyes Cody's face continues to reverberate in the American imaginary.[54]

The Crying Indian first aired in alignment with the second annual Earth Day, 1971. The American Ad Council's strategic timing aimed to fuse its message with the publicity and positivity surrounding Earth Day activities. "The moment [of the ad's promotional release] was auspicious," notes Shepard Krech in *The Ecological Indian*, since "between fifteen and twenty million people had assembled on Earth Day to create the groundswell for an Environmental movement unprecedented in scale and zeal."[55] The PSA opens with Iron Eyes Cody depicting a Native American steering a canoe downriver, shrouded in mist. As menacing music escalates, the camera pans out to reveal a heavily industrialized setting. A voice-over begins speaking in grave tones as Cody disembarks upon a shore strewn with litter: "Some people have a deep, abiding respect for the natural beauty that was once this country." The camera's focus moves off of Cody's profile to a busy four-lane highway in the background. As one car drives past, a passenger tosses a bag of fast food out the window as the voice-over says, "and some people don't." The bag bursts just in front of Cody's moccasins, scattering half-eaten burgers, fries, and debris across the concrete. Looking up from the debris, he turns his head and meets the camera's gaze as a single tear peels off onto his cheek. The voice-over says sternly, "People start pollution. People can stop it" (Fig. 3.1).

Iron Eyes Cody's single tear quickly became an icon of environmental concern and a poignant reminder for responsible waste management, successfully suturing its message to the developing Environmental movement. The spot garnered nationwide acclaim. *Ad Age* magazine ranks it as #50 in their "Top 100 Advertising Campaigns of the 20th Century," while *Entertainment Weekly* placed it in a list of the "50 greatest commercials for all time" and *TV Guide* ranked it 16th in their

Fig. 3.1 The Crying Indian, Keep America Beautiful (KAB), 1971. Reprinted with permission

"Fifty Greatest Commercials" list. Some twenty-seven years after the first appearance on screens across America, *The New York Times* referred to it as, "[o]ne of the best-known, most-honored commercials ever."[56] Marketing analyst Robert Cialdini claims "[it] was widely thought to be so moving that many consider it perhaps the most effective public service announcement of all time," often credited for bringing the subject of litter to the level of national attention.[57]

The din of praise for the Crying Indian PSA has muted its historical context and the motivations behind its creation. Once contextualized, it's clear the PSA's environmental ethic was crafted to deflect attention away from the corporate entities that profited from beverage bottling. More broadly, the Crying Indian PSA sought to contain the emerging environmentalism, which was mounting public pressure for holding corporate industries accountable for environmental damage. This context fundamentally shifts analysis of how the PSA functions rhetorically. The ad's pivotal timing, immediate impact, and enduring legacy are evidence for approaching it as a foundational text in the rise of eco-friendly rhetorics and their current pervasiveness within environmental discourse.

In the first few years of the 1950s, refillable glass bottles were beginning to be overtaken by aluminum cans as the container of choice for soft drinks and beer. Marketed as convenient "throwaways," these cans provided a sizable boost in sales for bottling retailers, since they could produce containers that had no deposit fee attached, as was the case with glass bottles. Nor did bottling companies have to concern themselves with the collection, cleaning, and reuse of their products. Cans were designed as one-time-use, which necessitated more production, resulting in higher profits. Though "disposable" cans were a boon to bottling and manufacturing industries, they quickly became the bane of local and state governments who dealt with the explosion of litter that followed (Fig. 3.2).

When the Vermont legislature began drafting a bill that banned non-reusable beer bottles, vested corporate leaders came together to respond to what they perceived as a threat to a revenue stream of grand potential. Recognizing that other states were looking to solve their own pollution issues—and that Vermont's legislation served as a clear model for curbing it at the source—these businessmen set to the task of persuading policy-makers and the public that pollution should be tackled in a different fashion. In 1953, the same year Vermont passed legislation outlawing non-refillable containers, a consortium of prominent

Fig. 3.2 1950s magazine advertisement for "throwaway" steel cans. Retrieved from https://www.ericwrobbel.com/collections/disposable-2.htm. Reproduced with permission

businesses, including Coca-Cola, Anheuser-Busch, and Philip Morris, joined together to found the nonprofit organization, Keep America Beautiful (KAB), with the goal of combatting roadside litter and to "promote a national cleanliness ethic."[58] The nation's first "bottle bill" was short-lived, however. Lasting only four years on the books, the bill was met with a "non-renewal" decision after aggressive lobbying against it by industry groups.[59] It would be another fifteen years before Vermont would pass another bottle bill.

By the late 1960s there existed palpable, mounting social and political pressure on corporations to account for their role in environmental degradation. In response, KAB mounted their largest campaign yet. Burson-Marsteller, "the global public relations firm famous for its list of clients with environment-related publicity problems," began work on the Crying Indian spot pro-bono for KAB.[60] Not coincidentally, they were also working for the American Can Company, one of the world's largest manufacturers of tin cans. As a campaign financed by beverage bottling corporations and other vested parties, the Crying Indian's constitutional aim was not to remedy the problem of pollution. Rather, the campaign was a calculated corporate effort to contain and redirect growing public pressure on manufacturers to stop producing non-refillable bottles. The goal was to lessen, if not wholly eliminate industry-accountability for creating the products that create pollution. The campaign strategy was to transfer blame and its execution pivots on collective guilt.

Scapegoat Analysis in Environmental Communication

"[T]he scapegoat," Kenneth Burke insists, "is a concentration of power."[61] It is the "chosen vessel" made to carry connotations of disgrace, disobedience, and disease.[62] By banishing this contaminated vessel, others are symbolically cleansed, able to return to the supposed default state of being orderly and whole. As a rhetorical mechanism for regulating social dynamics, the scapegoat has a lengthy history with myriad variations, though the principal function is consistent: achieve purification via relegation and separation.[63] For Burke, what makes scapegoating a "revolutionary kind of expression" and so "ritually gratifying" is its capacity to combine "in one figure contrary principles of identification and alienation."[64] The scapegoat, in other words, achieves a sense of unification through division, promising virtue and harmony for the many through the exclusion of the few. As its historical track record proves, scapegoating is a powerful process for mobilizing publics and shaping social structures.

Within the field of environmental communication, analyses of scapegoating have focused on how specific individuals get rhetorically cast as contaminated and are then socially isolated. In "Scapegoating in the Wild: A Burkean Analysis of Two Outdoor Adventures Gone Wrong," Molly Hartzog shows how Chris McCandless, whose journey into the Alaskan bush to live simply off the land ends tragically with his death, is ultimately "cast as a (literal and figurative) outsider to society."[65] In contrast, Aron Ralston, who, after getting pinned by a boulder and being trapped for five days in a canyon, survives by cutting off his arm, "emerges as a purified being … who has overcome their wrongdoings and [can thus be] reintegrated into society."[66] Hartzog's analysis shows how scapegoating techniques mobilize each wilderness event in service of deep-seated cultural narratives that prescribe how humans should properly relate to their environment.

Casey Schmitt also applies Burke's theories for analyzing the scapegoating of individuals within environmental contexts. Schmitt contends that an "emergent genre [of] *scapegoat ecology*" is being facilitated through social media networks, wherein "a public of environmentally minded individuals focuses attention and vitriol on a single person for being particularly harmful to the environment."[67] A ripe example is Walter Palmer, a dentist whose trophy hunting of a lion went viral, igniting a firestorm of hostility. "Scapegoat ecology is a form of character assassination," Schmitt argues, "that reduces its target to … nothing more than a symbol of cruelty or stupidity toward the natural world."[68] Character assassinations provide a feel-good, but ultimately false validation of one's own environmental ethic, Schmitt contends. Complex issues get reduced and attributed to a handful of offenders, producing a semblance of environmental stewardship without actually doing any of the hard work.

In arguing for scapegoat ecology as an emergent genre within environmental communication, Schmitt warns that a dangerous shift is afoot. What is at stake, he argues, is the loss of motivational incentive for meaningful action:

> [Scapegoat ecology] is a move away from the social responsibility campaigns of 1960s, 70s, and 80s environmentalism that stressed collaboration through a critical mass of *intentional*, *constructive*, *individual action*,

> reminding people to turn off light switches and tap water when not in use, urging them to "Give a Hoot" and not pollute, or insisting that "Only You Can Prevent Forest Fires."[69]

In focusing on a singular strain of scapegoating, Schmitt offers a productive account of how social media amplifies and accelerates the scapegoating process of egregious environmental offenders. In lamenting the turn away from campaigns anchored in individualized action, however, a far more pervasive and persuasive scapegoating process is overlooked. The brand of environmentalism that emphasizes "collaboration through a critical mass of *intentional, constructive, individual action*" was—and is still—cultivated in part by corporate interests seeking to transfer blame for pollution and to protect lucrative revenue streams.

Scapegoating, Recalibrated

If one applies Burke's theory of victimage to the PSA, its basic outline looks like this: bottling manufacturers and other vested corporations seek to "ritualistically cleanse themselves by loading the burden of their own iniquities" upon another entity, in this case the American consumer, who "performs the role of vicarious atonement."[70] The "iniquities" of creating environmentally toxic and wasteful materials in search of profit are transferred to the consumer, who is framed as responsible for the product's eventual impact. What makes this seemingly straightforward account so notable, however, is how the corporate interests "load the burden" onto the scapegoat, what gets transformed in the transfer, and how the guilt orients subsequent behavior.

The ad begins with Cody canoeing downriver, in his supposedly natural environment, symbolizing oneness with nature. As timpani drums and horns escalate a sense of tension, the camera pans out to reveal factories, towering cranes, and cargo containers. Cody is suddenly dwarfed, visually subordinated to the iconic forms of industrialization cast as threatening and unnatural. The juxtaposition of Cody's canoe set against a factory-scape invokes a dispiriting history of stolen land, colonization, and genocide. The first half of the ad invites disgust for—and also association with—a despoiled ecosystem stolen from indigenous peoples. The second half focuses on trash and its disposal. The ad adroitly links a sequence of historical allusions that align American viewers with a ponderous guilt for genocide and environmental devastation, fusing it to the problem of litter. The scapegoat's burden is not just roadside rubbish, but also the industrial and colonial forces ravaging the land and its inhabitants.

The PSA generates guilt by luring viewers into a conflicting set of identifications. We are encouraged to identify with Cody as the protagonist and feel sympathy for his circumstances. Viewers are also aligned with the colonizing culture that is the source of Cody's misfortune. The incongruity invites feelings of collective guilt. It is not necessary, after all, to participate directly in the harmful actions to experience guilty feelings for them. One must only feel aligned with the transgressor, whether it be a nation, a race, or an economic system. Indeed, what makes collective guilt so intriguing from a rhetorical perspective is that it arises primarily, and at times exclusively through one's identification with a group. This is also what makes it so exploitable.

From a Burkean perspective, lumping such a large swath of transgressions together as if they were equal and thereby collapsing distinctions between direct guilt and collective guilt is *mystification*. The PSA effectively constructs, as Burke would put it, "a fog of merger-terms" that conceals the motives of the corporate interests behind KAB. "If, when there is a quarrel over property," whether over its acquisition or in this instance, responsibility of disposal, "instead of confronting it squarely you begin … looking for remote kinds of metaphysical or theological anguish and alienation embedded in the very essence of humanity," Burke warns, "you are blinded by a principle of 'mystification'."[71] In this case, when the issue is who should be responsible for product packaging—those who produce it or those who touched it last—and you are confronted with the inherited collective guilt of an entire nation and its present sins resulting from a heavily industrialized means of production, you have been blinded by mystification. While you are pondering how to get rid of disquieting feelings, Burke says, "empires are striving for *world markets*."[72]

Consider how different the narrative would be if Cody, having waded through trash only to have a bag of food thrown at him, responds not with sadness, but with an equally reasonable emotion given the circumstances: anger. Anger tends to involve an elevated focus on the perceived *cause* of distress, and thus orients behavior toward retribution. Anger encourages one to investigate and evaluate justice and blame more closely, motivating action aimed at the source of the cause. Sadness tends to lead in the opposite direction—toward assessing the event's consequences for the victim. Empathetic sadness directs one toward the victim—Cody and all he symbolizes—but also orients one toward guilt, as viewers are simultaneously framed as being responsible for his condition. This guilt in turn breeds desire to "heal" the situation and expunge the negative feelings. The tear's rhetorical design, in short, orients one toward guilt and away from indignation—toward *reconciliation* and away from *retribution*.

Though Burke's theories offer a valuable toolkit for unpacking many scapegoat scenarios (as Hartzog and Schmitt demonstrate), the use of collective guilt in this instance requires some recalibration. For example, in Burke's scapegoating, the process for loading one's iniquities onto others is a threefold process:

> (1) an original state of merger, in that the iniquities are shared by both the iniquitous and their chosen vessel; (2) a principle of division, in that the elements shared in common are being ritualistically alienated; (3) a new principle merger, this time in the unification of those whose purified identity is defined in dialectical opposition to the sacrificial offering.[73]

Both corporations and individual consumers pollute and waste resources; they just do so on vastly different scales. The second step, division, executes the transfer of blame. In a traditional Burkean reading, this stage summons sharp distinctions between "them" and "me/us." By claiming that *they* are a source of impurity and are radically different from *me/us*, the stage is set for symbolic exile and cleansing. For example, Schmitt notes that the "right/wrong, good/bad, or smart/stupid distinctions required in scapegoating thrive on character attacks, like name calling, mockery, and expressions of disgust."[74] Although clear binaries are often invoked to create division, there is no evidence to indicate that it's required to achieve the principal function of scapegoating: transferring a disproportionate burden of blame onto a chosen vessel in order to absolve oneself.

In the Crying Indian PSA, the "principle of division" is enacted through an *inversion* of the sharp binaries we have come to expect from traditional scapegoating tactics. Unlike Burke's model, instead of drawing hard boundaries between "they" and "we" so that the many might be purified by expelling the contaminated few, the PSA's strategy actually aims at *broadening* the we. The ad's imagistic associations and emotional invitations work to align industrial pollution with the viewer, a pairing that's explicitly affirmed in the concluding syllogism: "People start pollution. People can stop it." *The sins of a few are loaded onto the backs of the many.*

In a typical scapegoating, the party seeking purification rhetorically distances itself, such that "people" would be defined in opposition to corporations. Yet here, the metonyms of industrial production are pronounced. The emblems of corporate pollution are shown—quite literally—in a bad light, with factory-scapes tinted in tones of polluted yellow and belching smokestacks silhouetted to appear menacing. The factories are instead aligned with the mass of citizens. They are toxic byproducts of the American people. The corporation as a separate actor effectively dissolves into "people," absolving themselves not through a *them* versus *us* dichotomy, but through a *diffusion* of both into a capacious *we* that becomes too conceptually dispersed to allow for more specific targets of blame.

The scapegoating strategy seen in the Crying Indian was not necessarily new. Rather, it was a crystallization of a rebuttal to emerging environmental activism delivered at a kairotic moment on a national level. Both environmental activists and pro-industry PR groups witnessed the substantial public response to Earth Day's launch in 1970 and saw quite clearly that environmentalism was poised to go mainstream. The debate shifted from *whether* we should respond to ecological destruction to a battle over framing *how* we should respond. As one activist puts it:

> It is the capitalists, not the public, who have computed the political logic of ecology….Their PR men have been working overtime promulgating [the frame that] [c]ars pollute the air because you want a car….Peasants are exploited in Guatemala because you insist on putting bananas on your Shredded Wheat….[This argument] deludes people into thinking their individual decisions can help, and it keeps people from recognizing that the only way they can really help is to act collectively.[75]

A 1970 *Newsweek* article, "The Ravaged Environment," provides a contrasting view:

> For the villain is not some profit-hungry industrialist who can be fined into submission, nor some lax public official who can be replaced. The villains are consumers who demand … new, more, faster, bigger, cheaper playthings without counting the cost in a dirtier, smellier, sicklier world.[76]

The effectiveness of the Crying Indian involves its timing, as well as its scapegoat mechanism. The corporate entities profiting from disposable packaging were indeed purified—they successfully deflected attention and vitiated public pressure. The scapegoat, however, was not one specific person or even a particular group. After obfuscating the proportionality of guilt, the blame is transferred to the "people," which functions as a synonym for individual consumers. The individual—*qua individual*—is the scapegoat, onto whose shoulders are loaded all the iniquities of industrial capitalism, by virtue of participation in the system. The blame is transferred through a *massive atomizing* of scapegoats. By instrumentalizing collective guilt in this manner, the PSA helped entrench an individual-focused approach to environmentalism for decades to come.

Redirecting Responsibility Through Consumer Complicity

In 2010, citizens of Plaquemines Parish, Louisiana gathered for a town hall meeting. In the oily wake of the BP/Deepwater Horizon spill, they assembled to receive updates on triage efforts and to demand guarantees that BP will be held accountable for the nearly 70,000 square miles of ocean directly impacted by the oil. Coast Guard Captain Ed Stanton, who led cleanup operations, responded to vociferous concerns regarding the chemical dispersants they were applying to keep the oil from spreading further. Expressing some irritation, Stanton indicated that there are no perfect options for cleaning up the spill: "It's name your poison. Look, we're all guilty in this rummage—if you drive a car, you own part of this spill."[77] This rejoinder mirrors the logic of the Crying Indian PSA: shift attention away from *retribution* and toward *reconciliation*, redirect volatile anger toward feelings of collective guilt.

Deflecting corporate culpability through rhetorics of consumer complicity has only become more refined since Iron Eyes Cody's single tear. BP has repeatedly relied on this strategy, including the Prudhoe Bay oil spill of 2006. To dispel public outrage, the company hired Peter Sandman, a well-known risk communication consultant. Sandman administered a five-point plan based on "outrage factors": (1) *Voluntary Risk*. If you are able to convince the public they have voluntarily played a role in the incident, externally imposed risks can be downplayed; (2) *Fair Risk*. "The benefits of your product or service might greatly outweigh the risks," Sandman suggests; (3) *Familiar Risk*. By making the risk feel ordinary, the fear of it is diminished; (4) *Risk Control*. "A company can project these messages: 'Butt out' and 'Stop worrying.'

But a public that has no control over a risk will do neither," he writes; and (5) *Acknowledge Risk.* Building credibility by acknowledging mistakes will be more effective than stonewalling.[78]

Though Sandman does not explicitly discuss collective guilt, it is essential to his model's success. The first step of his model entails drawing attention to one's role in the incident—the stoking of collective guilt—and frame it as a voluntary choice. Consumers' connection to the harm is highlighted again in the fourth point, this time by projecting a sense of control. Sandman's five-point plan invokes complicity on many levels, and in each case, postulates a self-determining individual governed by rational faculties. His model is not only premised upon this figuration of the individual, it *actively forefronts it* in order to dispel outrage. It advances the frame that every engagement within economic and political structures is voluntary—the result of deliberative reflection and intentional choice-making among several options. This underlying conception of the individual is what allows charges of complicity in systematic environmental destruction to appear logical.

Captain Stanton's remark also illustrates how the strategies actively employed by corporations are readily internalized by other members of the public. Consider his enthymematic claim: The guilty "we" in "we're all guilty" is expanded to include anyone who drives a car. The qualifying "if" that begins the last sentence implies that driving is unconditionally a voluntary option, while the 2nd-person address of "you" individualizes and intensifies the agency in that choice. The rhetorical sequence is clear: emphasize connection to the harm; establish complicity; emphasize that connection as voluntary; stoke collective guilt; redirect anger; transfer the blame onto the individual *qua individual.*

Though anger at BP and belief in their culpability are not necessarily overridden by this scapegoating, these rhetorics of consumer complicity do work to mitigate those feelings. Even when anger persists and one seeks retribution against BP, the double-bind of environmental activism often reveals itself, as Jenny Rice points out in *Distant Publics: Development Rhetoric and the Subject of Crisis.* She notes that while many called for a boycott on BP gas following the spill, "if we consider the crisis across incongruent and asymmetrical networks, we are likely to find that this action does not carry the true mark of sustainability."[79] Temporarily choosing another gas company does nothing to change the structural systems that led to the disaster. Nor does boycotting a BP gas station hurt the company, as stations are likely franchised by independent owners. Rice notes a certain insularity to the exercise as well, since even bigger oil spills occur elsewhere, yet outrage isn't mobilized for those international incidents. The calls to boycott, she writes, "failed to place the event within multiple networks, which ultimately calls for us to consider much more dramatic changes than where we fill our tanks."[80] This is precisely the point that must be reached. Getting there in earnest, however, is often frustrated by variations of the hypocrite's trap, the double-binds of activism, and scapegoating rhetorics, all of which rely on environmental guilt as an impelling and mystifying force. Emotional literacies that attend to collective, environmental guilt can shift how we identify, digest, and orient to feelings of consumer complicity which scapegoat the individual and veil the extractive systems that so clearly warrant dismantling.

Unfriending Scapegoat Ecologies

The bulk of this chapter explores ways in which charges of complicity and provocations of collective guilt shape environmental communication by steering attention to the individual. The forms in which they manifest can be used to contain action (such as when environmental advocates are chastised for their hypocrisy) or generate it (encouraging the purchase of eco-friendly products, for example). Bids to be environmentally friendly are so prosaic, so trite in their singular status, that to analyze a specific example without the proper context risks missing how such appeals function as a cumulative force.

The vast majority of these rhetorics, after all, are preposterously inane. Take a recent article listing the ways you can "green" your email signatures. Its opening pitch is sanguine: "Love this little planet? Want to live and breathe the sustainability/eco-friendly message? Good on you. When it comes to awareness of the environment and caring for it every little bit helps."[81] The recommendations range from cheeky reminders not to print emails ("Trees have feelings too, please don't print this!") to sharing your favorite eco-friendly tips ("Saving power can save the planet—turn off all computers, electronics and wall sockets at the end of each day.").[82] It's easy (and enjoyable) to make fun of such content for being so cheerily unaware, urging you to use your email footer to "get people in that eco-friendly mood with some environment-loving quotes" while living within the sixth mass extinction.[83]

On its own, this article is ludicrous. As a contributing element to one of the most pervasive forms of environmental communication, however, it takes on new meaning. The truism that in caring for the planet "every little bit helps" seems innocuous as a solitary statement, yet in the aggregate, it reinforces the logic that small, isolated changes made by individuals will add up to offset the systemic, substantial damage that occurs on a structural level each day. I recognize that such a strong claim appears absurd when attached to such a petty example. And that is precisely how eco-friendly rhetorics exert meaningful influence on environmental discourse. Their capacity to convince is owed "much more to trivial repetition and dull daily reinforcement than to exceptional rhetorical skill."[84] With environmentally friendly rhetorics, it actually is the case that every little bit helps—in that their persuasiveness is a result of accrual and pervasive circulation. Taken piecemeal, it is easy to dismiss their rhetorical import. Within our everyday lives, their ubiquity renders them a backdrop that fades out of focus. The combination of being perfunctory and pervasive conceals how eco-friendly rhetorics reinforce the assumption that incremental, individualized action is the default response toward ecological crisis.

Eco-friendly rhetorics animate scapegoat processes on an imperceptibly small scale, continually stoking low-lying levels of collective guilt for complicity in an ecologically destructive system. What is essential to note, however, is that they also offer a route for *immediate and achievable atonement*. By encouraging one to recycle their plastic bottle or purchase the more environmentally friendly dish soap, eco-friendly rhetorics also provide a means of absolution for the collective guilt they just agitated. Opportunities abound for purification via market-based solutions and consumption management techniques. "Instead of feeling guilty about the huge gaps between wealthy and poor, the ways consumerism causes global warming, or how our daily pleasures cause rain forest destruction and despoil the sea," Richard Wilk quips in *Green Consumption: The Global Rise of Eco-Chic*, "we can drink a few cups of fair-trade coffee and eat a rainforest crunch bar and instantly feel better."[85] Any relief from a mild sense of collective guilt gained through green consumerism, of course, is transitory. Eco-friendly rhetorics don't just alleviate guilt, they provoke it by design.

Moreover, the structural problem is not changed; in fact, it has been exacerbated, since "eco-conscious" consumerism has catalyzed a fresh wave of commodity development and spurred enormous growth.

As I noted at the chapter's outset, going green is big business—simultaneously a consequence of and cause for the prevalence of eco-friendly rhetorics. The loop of collective guilt and consumer atonement helps to further circulate eco-friendly rhetorics, perpetuating a kind of cruel optimism, in which "something you desire is actually an obstacle to your flourishing," as Lauren Berlant describes it.[86] As a principal means of reinforcing an individualistic orientation toward addressing ecological crisis, eco-friendly rhetorics maintain the conditions that ultimately undermine calls for meaningful change, all while they explicitly call for meaningful change.

That eco-friendly acts and products are not going to solve our ecological crisis is achingly obvious, including to many if not most people who try to be environmentally friendly. I am not attempting to reveal that these messages are used to make money in the capitalist system, which is glaringly self-evident. Rather, my aim has been to illustrate how this conspicuous strain of environmental communication functions rhetorically in ways that have gone unarticulated. It is tempting to cast eco-friendly rhetorics as little more than shallow marketing attempts or as an innocuous form of environmentalism, which, even if not the grand solution, is at least better than the non-eco-friendly alternative. My argument is that eco-friendly rhetorics, far from being innocuous, are a powerful means for framing environmental discourse around personal responsibility. They enact a unique form of scapegoating that subtly and systematically transfers the blame for environmental damage onto the individual, and in the process, further normalize the individual as independent and rational in their deliberate decisions to act voluntarily within a social and economic system that is wreaking havoc on earth's ecosystems.

Collective guilt is central to the success of eco-friendly scapegoating. Feelings of distress generated by recognizing one's complicity in a collective atrocity, however, are largely illegible with emotional literacies constructed to only read guilt as attributable to individuals and their direct actions. As seen with *No Impact Man*, even when one pushes these actions to the maximum, their limited range of impact and difficulty of implementation quickly becomes evident. Calls to dramatically change the system, however, are often met with accusations of hypocrisy. Such accusations direct one *back to the complicity* that helped spark that call to action in the first place. In cultivating an ecological perspective, options for alleviating guilt not only shrink, but opportunities for feeling *more* guilt expand.

Collective guilt need not be an individualizing force. Feelings of environmental guilt, after all, are recognitions of connection to the wider communities and ecologies within which we are enmeshed. Our environmental guilt is a rich and potent source for generating creative, reparative action for the ecological systems currently under siege. The desire for making amends that accompanies collective guilt cannot be fully satiated through individual actions, however. To transform environmental guilt's affective pull toward reconciliation, we must act as part of something greater than ourselves. Environmental guilt is a form of collective guilt, and collective guilt, I argue, calls for collective action.

Notes

1. Davis Guggenheim (Director) (2006) *An Inconvenient Truth: A Global Warning* [Motion picture]. Hollywood: Paramount.
2. Kenneth Burke, *A Rhetoric of Motives* (Berkeley: University of California Press, 1950): 26.
3. Nielson, "Consumers'-Goods Brands that Demonstrate Commitment to Sustainability Outperform Those That Don't," October 12, 2015, https://www.nielsen.com/us/en/press-room/2015/consumer-goods-brands-that-demonstrate-commitment-to-sustainability-outperform.html.
4. Ashlan Bonnell, "Consumer Attitudes Toward Green Brands Reach All-Time High," April 2, 2015, https://blog.marketresearch.com/sustainability-in-america-consumer-attitudes-toward-green-brands-reach-all-time-high.
5. Wendy Brown, *Undoing the Demos: Neoliberalism's Stealth Revolution* (Brooklyn: Zone Books, 2015): 49.
6. "Recognizing the Duty of the Federal Government to Create a Green New Deal," H.Res.109, 116th Congress (2019), https://www.congress.gov/bill/116th-congress/house-resolution/109/text.
7. Ibid.
8. Ibid.
9. Lukas Mikelionis, "2020 Democrats Jump to Endorse Green New Deal Despite Spending Hundreds of Thousands on Air Travel—Including Private Jets," *Fox News*, February 9, 2019, https://www.foxnews.com/politics/2020-democrats-jump-to-endorse-green-new-deal-despite-spending-hundreds-of-thousands-on-air-travel-including-private-jets.
10. Ibid.
11. Isabel Vincent and Melissa Klein, "Gas-Guzzling Car Rides Expose AOC's Hypocrisy," *New York Post*, March 2, 2019, https://nypost.com/2019/03/02/gas-guzzling-car-rides-expose-aocs-hypocrisy-amid-green-new-deal-pledge/.

12. Ibid.
13. Jen Schneider, Steve Schwarze, Peter Bsumek, and Jen Peeples, *Under Pressure: Coal Industry Rhetoric and Neoliberalism* (London: Palgrave Macmillan, 2016): 107.
14. Ibid., 113.
15. Ibid., 108.
16. Ibid., 118.
17. David Harvey, *A Brief History of Neoliberalism* (Oxford: Oxford University Press, 2005): 3.
18. Jason Read, "A Genealogy of Homo-economicus: Neoliberalism and the Production of Subjectivity," *Foucault Studies* 6, no. 1 (2009): 25.
19. The most succinct articulation of neoliberal subjectivity comes from the unlikely source of DJ Quik, who in 1995 rapped the lyric, "if it don't make dollars, it don't make sense." See "Dollaz & Sense," *Safe+Sound* (Skip Saylor Recording, 1995).
20. Schneider et al., "Hypocrite's Trap," 123.
21. Ibid., 124.
22. Ibid., 116.
23. Nathaniel Johnson, "Lessons from Boulder's Bad Breakup," *Grist*, January 19, 2018, https://grist.org/article/lessons-from-boulders-bad-breakup/.
24. Ibid., 116.
25. Colin Beavan, *No Impact Man: The Adventures of a Guilty Liberal Who Attempts to Save the Planet and The Discoveries He Makes About Himself and Our Way of Life in the Process* (New York: Farrar, Straus and Giroux, 2009): 14.
26. Ibid., 11.
27. Ibid., 36.
28. Lauren Berlant, "Thinking About Feeling Historical," *Emotion, Space, and Society* 1, no. 1 (2008): 6.
29. Ibid., 6.
30. Ben Highmore, *Cultural Feelings: Mood, Mediation, and Cultural Politics* (London: Routledge, 2017): 50.
31. Elizabeth Kolbert, "Green Like Me: Living Without a Fridge and Other Experiments in Environmentalism," *The New Yorker*, August 31, 2009, https://www.newyorker.com/magazine/2009/08/31/green-like-me.
32. Ibid.
33. Mark Jenkins, "Radical Change Writ Small, For the Planet's Sake," *NPR*, September 3, 2009, https://www.npr.org/2009/09/03/112447407/radical-change-writ-small-for-the-planets-sake.
34. Alexandra Jacobs, "Green Life," *The New York Times*, October 15, 2009, https://www.nytimes.com/2009/10/18/books/review/Jacobs-t.html.

35. Jen Schneider and Glen Miller, "The Impact of 'No Impact Man': Alternative Hedonism as Environmental Appeal," *Environmental Communication* 5, no. 4 (2011): 477.
36. Chris Russill, "Steve Schneider and the Double-Ethical Bind of Climate Change Communication," *Bulletin of Science, Technology, and Society* 30, no. 1 (2010): 61.
37. Anthony Chaney, *Runaway: Gregory Bateson, the Double Bind, and the Rise of Ecological Consciousness* (Chapel Hill: University of North Carolina Press, 2017): 5.
38. Ibid., 6.
39. Schneider et al., "Hypocrite's Trap," 125. The final sentence has been abridged, excluding the clause: "… far removed from the 'common sense' deliberations of the industry."
40. Laura Gabbert and Justin Schein (Directors) (2010) *No Impact Man* [Motion picture]. United States: Oscilloscope Laboratories.
41. Marilyn DeLaure, "Environmental Comedy: *No Impact Man* and the Performance of Green Identity," *Environmental Communication* 5, no. 4 (2011): 453, 458.
42. Schneider and Miller, "The Impact of 'No Impact Man'," 485.
43. Mary Ellen Shoup, "Bottled Water Surpasses Soda in Consumption with 86% Purchase Rate Among Americans," *Beveragedaily.com*, March 27, 2017, https://www.beveragedaily.com/Article/2017/03/13/Bottled-water-surpasses-soda-in-consumption-for-the-first-time.
44. Sandra Laville and Matthew Taylor, "A Million Bottles a Minute: World's Plastic Binge 'As Dangerous as Climate Change'," *The Guardian*, June 28, 2017, https://www.theguardian.com/environment/2017/jun/28/a-million-a-minute-worlds-plastic-bottle-binge-as-dangerous-as-climate-change.
45. Ibid.
46. Laura Parker, "Here's How Much Plastic Trash Is Littering the Earth," *National Geographic*, December 20, 2018, https://news.nationalgeographic.com/2017/07/plastic-produced-recycling-waste-ocean-trash-debris-environment/.
47. Timothy Luke, *Ecocritique: Contesting the Politics of Nature, Economy, and Culture* (Minneapolis: University of Minnesota Press, 1997): 118.
48. See Rachel Carson, *Silent Spring* (Boston: Houghton Mifflin, 1962).
49. Luke, *Ecocritique*, 118.
50. I follow Michael Calvin McGee in defining social movements in fundamentally rhetorical terms—patterns of meaning, suasion, belief, and behavior—and not as a collection of people.
51. Michael McCloskey, Sierra Club records, carton 136, December 23, 1970.

52. *Focus Midwest* 8, no. 54 (1971): 48.
53. Luke, *Ecocritique*, 118.
54. Examples of the PSA's enduring recognition include its parody in the *The Simpsons* 200th episode, an allusion in *World of Warcraft*, and references in numerous films including *Wayne's World 2* and *Kingpin*, as well as in television sitcoms such as *Friends* and *Married…with Children*.
55. Shepard Krech, *The Ecological Indian: Myth and History* (New York: Norton, 1999): 15.
56. Stuart Elliott, "An Environmental Campaign is Back by Popular Neglect," *The New York Times*, April 22, 1998.
57. Robert Cialdini, Noah Goldstein, and Steve J. Martin, *Yes!: 50 Scientifically Proven Ways to be Persuasive* (New York: Simon and Schuster, 2008): 19.
58. "About Us: Mission and History," *KAB.org*, https://www.kab.org/about-us/mission-history.
59. Finn Arne Jørgensen, "A Pocket History of Bottle Recycling," *The Atlantic*, February 27, 2013, https://www.theatlantic.com/technology/archive/2013/02/a-pocket-history-of-bottle-recycling/273575/.
60. Ginger Strand, "The Crying Indian," *Orion Magazine*, Nov/Dec 2008, http://www.orionmagazine.org/index.php/articles/article/3642/.
61. Kenneth Burke, *A Grammar of Motives* (Berkeley: University of California Press, 1962): 407.
62. Ibid., 406.
63. The word "scapegoat" first appears in William Tyndale's 1538 translation of the Bible. Tyndale's use of "scape" is a case of aphesis—dropping an unstressed initial vowel—and refers to "escape," indicating that the goat was given the opportunity to flee into the wild.
64. Kenneth Burke, *A Rhetoric of Motives* (Berkeley: University of California Press, 1962): 140–141.
65. Molly Hartzog, "Scapegoating in the Wild: A Burkean Analysis of Two Outdoor Adventures Gone Wrong," *Environmental Communication* 9, no. 4 (2015): 535.
66. Ibid.
67. Casey R. Schmitt, "Scapegoat Ecology: Blame, Exoneration, and an Emergent Genre in Environmentalist Discourse," *Environmental Communication* 13, no. 2 (2019): 153.
68. Casey R. Schmitt, "The Dentist Who Kill Cecil the Lion, and a Lesson in Scapegoat Ecology," *Routledge Handbook of Character Assassination and Reputation Management*, eds. S. Samoilenko, E. Shiraev, M. Icks, and J. Keohane (New York: Routledge, forthcoming).
69. Ibid., 154.
70. Kenneth Burke, *Permanence and Change: An Anatomy of Purpose*, 3rd ed. (Berkeley: University of California Press, 1984): 271.

71. Kenneth Burke, *Rhetoric of Motives* (Berkeley: University of California Press, 1969): 108.
72. Ibid.
73. Burke, *Grammar*, 406.
74. Schmitt, "Scapegoat Ecology," 155.
75. Gene Marine, "Scorecard on the Environment," *Ramparts*, December 20, 1973.
76. Kenneth Auchincloss, "The Ravaged Environment," *Newsweek*, January 20, 1970, 32.
77. Kathy Lohr, "In Louisiana, BP Tries to Quell Fears Over Spill," *National Public Radio*, May 13, 2010, https://www.npr.org/templates/story/story.php?storyId=126805557.
78. Nancy Kennedy, "Outrage-ous," *Shield: The International Magazine of BP Amoco*, 1992, http://www.psandman.com/articles/shield.htm.
79. Jenny Rice, *Distant Publics: Development Rhetoric and the Subject of Crisis* (Pittsburgh: University of Pittsburgh Press, 2012): 43.
80. Ibid.
81. Lisa Tate, "5 Little Ways to Make Your Email Signatures Green," July 6, 2018, https://www.crossware.co.nz/blog/5-ways-to-make-your-email-signatures-green/.
82. Ibid.
83. Ibid.
84. Burke, *Rhetoric of Motives*, 26.
85. Richard Wilk, "Foreward," *Green Consumption: The Global Rise of Eco-Chic*, eds. Bart Barendregt and Jaffe Rivke (London: Bloomsbury Academic, 2014): xv.
86. Lauren Berlant, *Cruel Optimism* (Durham: Duke University Press, 2011): 1.

CHAPTER 4

Guilty of Shame in the *Anthropocene*

Abstract This chapter asks how the key term *Anthropocene* rhetorically orients one toward ecological crisis and how it functions as a guide for collective belief and behavior. I argue that by attributing ecological abuse to humans as a species, the term activates an emotional framework of shame rather than guilt, which undermines rhetorical efforts to mobilize collective action aimed at fighting environmental harm. With guilt, culpability is fixed on a specific action; in shame, culpability is embedded in an attribute, which constrains one's capacity to redress the wrong.

Keywords Emotional ecologies · Guilt · Shame · Anthropocene · Ideograph

Although atmospheric chemist Paul Crutzen didn't invent the term *Anthropocene*, he was the first to advocate for its adoption in the geologic sciences because, as he noted in 2002, "in many ways [a] human-dominated, geological epoch [has begun]."[1] Scientist Will Steffen similarly argues that humanity has entered a new era, and *Anthropocene* appropriately labels how "the human imprint on the global environment has now become so large and active, that it rivals some of the great forces of Nature in its impact on the functioning of the Earth system."[2] Despite its still-pending status before the International Commission on Stratigraphy as an official title, many scientists around the world

T. Jensen, *Ecologies of Guilt in Environmental Rhetorics*,
Palgrave Studies in Media and Environmental Communication,
https://doi.org/10.1007/978-3-030-05651-3_4

have nevertheless adopted *Anthropocene* as the name for our current geological epoch. Nor has the lack of institutional validation hindered a more widespread embrace of the term as "a rallying point for geologists, ecologists, climate and Earth system specialists, historians, philosophers, social scientists, ordinary citizens and ecological movements, as a way of conceiving this age in which humanity has become a major geological force."[3] The term is now readily invoked in journalism (e.g., *The New York Times* piece, "The Age of the Anthropocene: Should We Worry?"[4]), modern art (photographer Edward Burtynsky's *Anthropocene Project*[5]), contemporary critical theory (Bruno Latour's "Agency at the time of the Anthropocene"[6]), and in rhetorical theory (Richard Grusin's *Anthropocene Feminism*[7] and Amy Propen's *Visualizing Posthuman Conservation in the Age of the Anthropocene*[8]). Crutzen may have introduced it as a scientific term at the turn of the twenty-first century, but *Anthropocene* now circulates widely, well beyond the boundaries of academe's Earth Sciences.

As a key term around which contemporary environmental discourses and perspectives develop, *Anthropocene* has a certain unifying effect, joining together varied spheres of science, humanities scholarship, and activism under its banner. Its expanded circulation has consequently expanded its meaning and rhetorical function. "The Anthropocene is, in important ways, a slogan for the age of climate change," writes Jedediah Purdy, environmental and constitutional law professor at Duke University.[9] Any definition of the term, he argues, "[is] not a statement of fact as much as a way of *organizing* facts to highlight a certain importance that they carry."[10] *Anthropocene*, in other words, hasn't so much been adopted as a scientific label for a geological epoch as it has been *adapted* for use across broader contexts. According to Purdy,

> "the Anthropocene" is an attempt to do what the concept of "the environment" did in the 1960s and early 1970s: join problems as disparate as extinction, sprawl, litter, national-parks policy, and atomic fallout into a single challenge called "the environmental crisis." Such a term is pragmatic: it tries to help people to act by gather the elements of their predicament together in a tractable way. The environment had to be named before people could join together to try and save it. The Anthropocene has to be named before people can try to take responsibility for it. Talking about the Anthropocene, then, can be a call to take responsibility for a changing planet.[11]

Anthropocene's pervasive uptake, Purdy suggests, is driven by its potential to be both motivational catalyst and shared focal point. That is, the term's rapid ascendance is propelled by a desire to move others to action: it helps them first recognize the problem, then assume responsibility for it, and finally, take steps to address the problem, both individually and collectively.

Animating *Anthropocene*'s growing circulation is a rhetorical venture—that the *name* is also a generative *frame*, capable of galvanizing collective will toward taking responsibility and thereby spurring ecological restoration. *Anthropocene* names an era, yes, but its persuasive power is derived from its capacity to organize a complex set of events into a conceptual framework that "in a tractable way" guides thinking and behavior in ecologically beneficial channels. How, then, this chapter asks, does *Anthropocene* rhetorically orient us toward ecological crisis? In calling for us to take responsibility for ecocide, what emotional orientations does the term invite? The vocabulary we use to describe something as colossal and imperative as the sixth mass extinction affects how we perceive the problem—and by extension, how we perceive our capacities in contributing to their solution.

I approach *Anthropocene* not as a geologic classification, but as a constitutive force of contemporary environmental rhetorics. My analysis attends to how the term functions *ideographically*, as a "high-order abstraction representing collective commitment to a particular but equivocal and ill-defined normative goal."[12] As a key term that acts as an affective catalyst and guide for emotional orientation, *Anthropocene* helps condition collective belief and behavior, simultaneously composing and reflecting social and political norms. I argue that *Anthropocene*'s rhetorical power pivots on its culpatory connotations, and that by attributing ecological abuse to humans in toto—as a species—the term directs attention away from specific actors and actions and, rather, shifts focus to some innate trait of humanity. Although *Anthropocene* has brought attention to pressing environmental issues, I argue the term activates an emotional framework of shame rather than guilt, which undermines an ecological perspective.

Contentions and Contenders of *Anthropocene*

In *Dark Ecology*, Timothy Morton bemoans that "[n]ot a day goes by recently without some humanities scholars becoming quite exercised about the term *Anthropocene*."[13] Morton then proceeds to get quite exercised himself, first cataloguing various charges levied against the designation, then positing rebuttals in explaining "[h]ow I learned to stop worrying and love the term '*Anthropocene*'."[14] The criticisms of *Anthropocene* are numerous and the debates volatile, as Morton's *Dr. Strangelove* allusion implies. Critics have argued the appellation is colonialist, racist, anthropocentric, hubristic, and a "blatant example of speciesism."[15] Each of these critiques has unique merits, though my focus here is on what unites them: contentions over who is included and what is elided by the "human" that etymologically and conceptually anchors *Anthropocene*. Furthermore, although there is fervent disagreement over the term's meaning, uses, and implications for engaging ecological crisis, the affective intensity of these debates signals agreement that *Anthropocene* has considerable power and potential for shaping contemporary environmental discourse. What is largely missing from these discussions, however, is consideration of how the term orients us emotionally to engage with culpability in a collective context.

Etymologically, the neologism *Anthropocene* is fundamentally ascriptive. It brings the Greek *anthropos* (ἄνθρωπος), meaning "human," together with *-cene*, derived from *kainos* (καινός), meaning "new" or "now"—in this case, a new era in earth's biospheric evolution, with "human" as its author. For Donna Haraway, this centering of "human" sets us up for some tired and uphelpful storytelling precisely at the moment when we sorely need "systemic stories of the linked metabolism, articulations, or coproductions ...of economies and ecologies," stories that are "relentlessly relational, sympoietic, and consequential."[16] Haraway laments the *Anthropocene*, "as a tool, story, or epoch to think with" on numerous grounds, as "[the] myth system associated with the Anthropos is a setup, and the stories end badly."[17] You may be familiar with this one, she jests darkly:

> The story of Species Man as the agent of the Anthropocene is an almost laughable rerun of the great phallic humanizing and modernizing Adventure, where man, made in the image of a vanished god, takes on superpowers in his secular-sacred ascent, only to end in tragic detumescence, once again. Autopoietic, self-making man came down once again, this time in tragic system failure, turning biodiverse ecosystems into flipped-out deserts...[18]

For Haraway, *Anthropocene* is not just the title of a bad story, it also serves as a euphemism for what might more accurately be called the

Capitalocene, a term that fixes the accusatory gaze on capitalism, with its fundamental premise of infinite growth on a planet with finite resources. *Capitalocene* directs our attention to the structural systems that facilitate ecological exploitation rather than human agents. Conceding that *Anthropocene* is "already well entrenched" and unlikely to be swapped out with *Capitalocene*, Haraway resigns herself to use *Anthropocene* sparingly and focus attention instead on a reparative alternative.[19] The stories prefigured by *anthropos* "must give way to geostories," she insists, which would forefront ecology—and not from the starting point of Man's place in ecology.[20] Haraway proposes "Chthulucene" as contrastive replacement, bringing together the Greek root of *khthôn* (χθών), meaning "earth," together with the now-familiar *-cene*. With *Chthulucene* we are oriented toward "ongoing multispecies stories and practices of becoming-with" and "[u]nlike the dominant dramas of Anthropocene and Capitalocene discourse, human beings are not the only important actors … [and] the biotic and abiotic powers of this earth are the main story."[21] *Chthulucene* names what humans should be paying attention to and using as a guide for our actions, making it more aspirational and less accusatory than *Anthropocene*.

Others share Haraway's critiques and have rebuked *Anthropocene* for what is minimized by the term's focalization on humans and how that orients us toward ecological disaster.[22] Derrick Jensen echoes Haraway in arguing that the term "contributes directly to the problems it purports to address," in that it helps to "naturalize the murder of the planet by pretending that the problem is 'man' and not this particular culture."[23] The sweeping ascription to "man" writ large reveals the same pattern of narcissism that has helped fuel ecocide—namely, it presents the behavior of the dominant culture "as representing 'man' as a whole," erasing in the process all indigenous cultures that have lived in far more sustainable ways.[24] "Humans aren't the ones 'transforming'—read: killing—the planet. Civilized humans are. There's a difference," Jensen asserts.[25] *Anthropocene* "manifests the supreme narcissism" of a colonial perspective in its all-encompassing attribution of culpability, rehearsing a well-worn Western practice of not acknowledging the existence, let alone worth of other cultures.[26] Though the term is circulated as a call to take accountability for environmental harm and spur a shift to sustainable ways of living, it orients one away from the indigenous cultures who have lived on the land for thousands of years, eliding the people, their systematic extirpation under the same colonial mindset, and the ecological knowledges they accrued over millennia.

Although *Anthropocene* linguistically centers humans, some argue that ultimately, the term *decenters* humans by spurring attunement to how we're ecologically connected. "Humanity is not at the center of the picture of the Anthropocene, opposing, by its powers of mind, the passive matter that encircles it," Jeremy Davies writes in *The Birth of the Anthropocene*.[27] "Instead," he adds, "human societies are themselves constructed from a web of relationships between human beings, non-human animals, plants, metals, and so on."[28] Davies contends that the larger discursive contexts in which *Anthropocene* circulates promotes this ecological attunement, which configure "humans" in a non-hierarchical schema. Consequently, he suggests, the term helps orients us toward radically interdependent relationships.

Jedediah Purdy takes this decentering potential even further. "Using the portmanteau term 'Anthropocene' for all these phenomena [related to ecological disaster]," he writes, "is willful, an effort to meld them into a single situation, gathered under a single name."[29] Out of this amalgamation comes a powerful proposition: "The revolution in ideas that the Anthropocene represents [is] *the end of the division between people and nature*."[30] For both Davies and Purdy, *Anthropocene* is human-centered in name only. The conceptual carrying capacity of the term exceeds its etymological constraints, they suggest, and its aspirational orientation—realization of ecological enmeshment—is ultimately more compelling than its culpatory centering of *human* as the responsible agent. In this way, their readings of *Anthropocene* echo Haraway's rationale for *Chthulucene*: both are focused on how a powerful key term can call an ecological identity into being, inciting a sense of collective belonging. This sense of connection will—ideally, ultimately—also generate a corresponding sense of responsibility and agency to take effective action.

Anthropocene locates "human" as a responsible agent of planetary change, but as these divergent views demonstrate, the interpretative range for what is connoted by the *Anthro* in *Anthropocene* can be expansive. Whereas Purdy and Davies see the potential for *Anthropocene* to kindle ecological attunement, Haraway and Jensen, among others, argue that it's more likely to replicate the human-centered rhetorics that have helped facilitate ecocidal behavior. The ambiguity of *Anthro* allows for a capacious construction of human, and this definitional imprecision, I argue, contributes to the ever-increasing circulation of *Anthropocene* and by extension, its persuasive power. That ambiguity, however, is fueled less by etymological root prefixes and more by the challenges of learning to think at the scaled-up level of *species.* In turn, thinking of culpability and responsibility at the level of species elicits complicated emotional orientations. Although debates on the merits of *Anthropocene* as a catalyzing, unifying key term have been ardent and the analyses abundant, to understand how *Anthropocene* helps structure contemporary environmental rhetorics, we must examine how it directs emotional attention and functions as a guide for collective belief and behavior.

Entanglements of Guilt and Shame

The origin story and etymology of *Anthropocene* are rooted quite plainly in ascribing culpability for global ecological transformation to humans, and its circulation is likewise premised upon this attributive character, such that it prompts routine acknowledgement of that collective blame. The emotional logic undergirding much of the *Anthropocene*'s use is that by recognizing one's culpability—however partial, and whether for doing wrong or failing to do right—a sense of guilt and responsibility will arise, which will in turn, animate urges to make amends. As Purdy reminds us, the "Anthropocene has to be named before people can take responsibility for it."[31] Within this sequence of emotional logic, however, there is an unreliable connection. The supposition that *Anthropocene* will generate guilt that leads to curative, redemptive action is specious.

Both shame and guilt, after all, are emotional acknowledgements of wrongdoing.[32] Distinctions between the two are critical, however. Donald Nathanson succinct summarizes key differences, noting, "[w]hereas guilt refers to punishment for wrongdoing, for violation of some sort of rule or internal law, shame is about some quality of the self. Guilt implies action, while shame implies that some quality of the self has been brought into question."[33] To consider how each emotion orients one's response to wrongdoing, we must first look to how they orient one toward the perceived *source* of that transgression. With guilt, culpability is fixed on a specific action; in shame, culpability is embedded in an attribute. This seemingly small difference has substantial ramifications, since attributes cannot be changed—only hidden—and while actions cannot be undone, they can be atoned for.

The common response to feelings of shame is embedded in the word itself: the etymology of shame is traced back to the verbal phrase, "to cover."[34] Shame's suasive forces urge one to retreat from view, shrink back from encounters with the public eye. This urge to hide from others is an enduring characterization of shame, as Charles Darwin's eloquent account from 1896 suggests. He observes, "Under a keen sense of shame there is a strong desire for concealment. We turn away the whole body, more especially the shame, which we endeavor in some manner to hide. An ashamed person can hardly endure to meet the gaze of those present."[35] In shame, you feel bad for the wrongdoing you have caused, but because the transgression is perceived as the consequence of who you are—rather than what you did—the tendency is to shield yourself from the judgment of others. That gaze of shame we seek to avoid, however, can be internalized, as we imagine how others might view us. Sara Ahmed notes that, "[c]rucially, the individuation of shame—the way it turns the self against and towards the self—can be linked precisely to the inter-corporeality and sociality of shame experiences." The individual and collective are co-extensively shaped through shame, just as they are through guilt.

Failing to recognize or appreciate the distinctions between guilt and shame is understandable, even justifiable, to a degree. The two are deeply entwined and share several core qualities. Their effects upon the body's sensorium are often indistinguishable, making physiological cues insufficiently reliable for differentiation. Moreover, contemporary education in emotional literacy is such that distinctions between the shame and guilt—and why those distinctions matter—are mostly neglected. We should also note that when we're in the thick of feeling badly, conditions are not exactly conducive to parsing apart whether we're experiencing guilt or shame.

Within rhetoric studies, discussions of guilt and shame have been obscured, if not wholly obstructed, by the fact that one of the enduring authorities on how *pathē* function persuasively—Aristotle—never mentions guilt once, and yet characterizes shame in ways that roughly correspond with our conceptions of guilt. This is not an oversight on Aristotle's part; there was simply no Greek word for guilt, because the concept had yet to emerge out of initial theorizations of shame.[36] Within classical philosophy, guilt is conceptually extracted from shame, a nativity that further justifies some of the continued confusion around their finer distinctions at a broader cultural level. However, that guilt was developed as a separate emotional category also indicates the need to differentiate its emotional trajectories from shame, clarifying the expression and examination of emotional experiences. It also demonstrates the cultural contingency inherent to the classification and theorization of emotion, which serves as a reminder that our emotional vocabulary is intimately connected with our capacities to identify, discuss, and contemplate patterns of feeling.

Despite their tangled evolutions and the difficulty in differentiating shame from guilt, the stakes for doing so are considerable. How we emotionally register our perceived wrongs can have profound effects on our capacity to respond. Research in behavioral psychology demonstrates that an abiding sense of shame is often far more damaging than guilt. Herant Katchadourian catalogues these contrasts:

> Shame leads one to deny, hide, and escape from a shame-inducing situation; guilt motivates one to take reparative action to undo the damage. Guilt enhances empathy toward others; shame disrupts it. Shame is likely to lead to anger, hostility, and blaming others; guilt leads to constructive responses to anger. A proneness to shame is associated with a wide variety of psychological symptoms (including low self-esteem, depression, anxiety) and conditions such as eating disorders and post-traumatic stress disorders. On the other hand, guilt is more effective than shame in restraining people from engaging in risky, illegal, and immoral behavior. Thus guilt is more "moral" an emotion than shame.[37]

The differences between shame and guilt are stylistically heightened in Katchadourian's accounting, which may leave the impression that the two can be neatly separated. However, the lived, daily navigation of challenging emotions—particularly those regarding mass extinction and our climate crisis—does not readily lend itself to such clean categorization. Moreover, the entanglements of guilt and shame become more knotted as we consider them in large-scale collective contexts. Nevertheless, there are consequences for developing a shame-based orientation toward wrongdoing in contrast to guilt, and whether *Anthropocene* tends to invite shame or guilt in framing environmental rhetorics, especially those seeking to mobilize effective action, is not a trivial matter. As it continues to become a discursive rallying point, the term impacts how we comprehend culpability for ecological harm on multiple levels—particularly, I argue, on our affective and emotional registers.

Species Shaming

"*Anthropocene* is about humans," Timothy Morton states, a statement that is at once both obvious and deceptively difficult.[38] "Consider that the term [*Anthropocene*] deploys the concept *species* as something unconscious, never totally explicit," he adds, pointing out that the word nevertheless connotes a totality—humans holistically conceived, "a mess of lungs and bacterial microbiomes and nonhuman ancestors … along with their agents such as cows and factories and thoughts …."[39] The real problem, as Morton sees it, is not that *Anthropocene* places human at the center (and therefore be dangerously anthropocentric, hubristic, or colonialist), but that we're utterly ill-equipped to conceive of "human" at the scale of ecology, which is precisely what's needed during this time of climate collapse and sweeping extinction.

Prevailing modes of thought tend to construct the human species in essentialized terms, as something that is "constantly, metaphysically present."[40] This commonplace view of species pivots on some essence—an intrinsic quality, woven into us as a collective—that makes us immutably human. "But *human* need not be something that is ontically given," Morton argues, as there "is no obvious, constantly present positive content to the human."[41] The term *Anthropocene* conjures human-as-species as the culprit for ecological upheaval, but it only becomes problematically anthropocentric when this essentialized notion of species is applied. "The concept of species, upgraded from the absurd teleological and metaphysical versions of old, isn't anthropocentric at all," he contends.[42] In fact, it's the complete opposite: "'*Anthropocene*' *is the first fully antianthropocentric concept*….because it enables us to think the human species not as an ontically given thing I can point to, but as a hyperobject that is real yet inaccessible."[43] The problem isn't the term—the problem is our antiquated, notion of human-as-separate-species, as set apart. For Morton, *Anthropocene* offers a chance to rethink the category of human "as something like the aurora, a mysterious yet distinct, sparkling entity."[44] It is an opening to reframe human in a more ecologically expansive manner.

Morton's account is not necessarily inaccurate, though it is manifestly inadequate. While *Anthropocene* may rightly serve as a means to emend how we conceive of the human species as a construct, Morton fails to sufficiently consider the audience for whom this is a plausible prospect. That is, he does not acknowledge the rhetorical impact that *Anthropocene* is *currently* exerting in how it orients a substantial number of people toward the most pressing ecological issues. It is true that *human* need not be something that is ontically given. It is also true, however, by Morton's own admission, that many do think of the human species in just this way. *Anthropocene* tempts this essentialized notion of *human*, and by accrediting planetary transformation to the species, it effectively attributes that misconduct to an intrinsic quality, some embedded element of what it means to be human. Because it directs attention to some holistic, constitutive quality of the species, *Anthropocene* impels a shame-based framework for engaging with global environmental harm. The term implies that climate collapse is a consequence of *who we are*, rather than *what we did*, leading attention away from specific acts and toward some awful aspect of our collective character.

But it is critical to note that we are not in the realm of human biology when engaging with the term *Anthropocene*—we are in the rhetorical realm. There is nothing in human DNA that indelibly links the species to an inevitable drive to destroying the ecologies that give it life. Identifying with the *anthropos* in *Anthropocene* is a rhetorical decision, one that has consequences for how we conceive of—and therefore respond to—ecological destruction. For most, *Anthropocene* is less philosophical invitation to rethink the entire category of human-as-species and more rhetorical force aimed at eliciting a sense of culpability in hopes of motivating a critical mass toward action. When those sensations of culpability are framed as a result of being human—rather than the actions of a particular culture—the term acts more as a shaming device than a guilt trip.

Anthropocene's species-centric rhetoric is buttressed by the prevalence of terms like "*man-made* climate change" and "*human-caused* climate change." Together, these terms create conditions that foster shame, rather than guilt, which have markedly different trajectories, even if phenomenologically they feel similar. Research in behavioral psychology has, for instance, "shown that shame-prone individuals are vulnerable to a variety of difficulties….Because the focus of shame is on the defective self, this painful emotion also has the effect of impairing empathy [whereas] shame-free guilt is positively correlated with adaptive characteristics, such as enhanced empathy."[45] The same study notes that "guilt-prone individuals are inclined to take responsibility for their actions, rather than to deflect blame onto others or onto elements of the situation."[46] As *Anthropocene*'s circulation continues to increase and the term is tacitly ratified as a "rallying point" for a diversity of groups, we must be vigilant for how it engenders emotional orientations that inhibit momentum toward effective action.

IDEOGRAPHIC *ANTHROPOCENE*

In a formal sense, *Anthropocene* labels a geologic era, but its persuasive power derives from its use as a framing device. As Purdy puts it, *Anthropocene* "[is] not a statement of fact as much as a way of *organizing* facts to highlight a certain importance that they carry….it tries to help people to act by gathering the elements of their predicament together in a tractable way."[47] In this way, the term functions *ideographically*—as "a high-order abstraction representing collective commitment to a particular but equivocal and ill-defined normative goal."[48] Reading *Anthropocene* through the lens of the ideograph further illuminates how the term operates rhetorically, persuading primarily through affective force and emotional orientation.

The grounding premise of Michael Calvin McGee's concept of the ideograph is that "[h]uman beings are 'conditioned,' not directly to belief and behavior, but to a vocabulary of concepts that function as guides, warrants, reasons, or excuses for behavior and belief."[49] The ideograph's ability to motivate group identification and reinforce collective commitment is its definitive function. Their persuasive power is the chief reason why rhetoricians have made frequent use of the ideograph as an analytical lens. Despite the myriad applications, most analyses privilege an ideograph's historical record or what McGee calls its "diachronic" component: how the ideograph has expanded and contracted in meaning and use over time.[50] How *Anthropocene* evolved from a technical term proposed by geologists to rallying point for action on ecological crises is traced earlier in this chapter, along with analysis of debates on its various meanings. Even the most complete diachronic analysis, however, "leaves little but an exhaustive lexicon understood etymologically," McGee warns, "and no ideally precise explanation of how ideographs function *presently*."[51] Understanding how *Anthropocene* influences belief and behavior in the present requires analysis of its *synchronic* dimension.

Drawing from Kenneth Burke's terminology, McGee aligns diachronic analysis with a "grammar" of public motive, while a synchronic reading seeks to capture its "rhetoric."[52] The latter attunes to the animating energy of ideographs, for "in real discourse, and in public consciousness, [ideographs] are *forces*."[53] For McGee, rhetoric "is the social equivalent of a verb in a sentence," and as such, appreciating how a word like *Anthropocene* actually moves people to action is a critical target for rhetorical analysis.[54] Apart from a few scant sentences, however, McGee offers little guidance on how to analyze an ideograph's rhetorical force and how it works to cohere and mobilize collectives. Kevin DeLuca rightly notes that the "neglect of the synchronic structure" in rhetorical criticism, "is not accidental but rather points to a certain lacuna in McGee's theorizing of ideographs."[55] Augmenting ideographic analysis with insights from affect theory, I argue, strengthens the ideograph's explanatory power, increasing its usefulness as a rhetorical tool for analyzing how language brings collectives into being and mobilizes bodies toward action.

Key terms like *Anthropocene*, Sara Ahmed argues, "increase in affective value as an effect of the movement between signs: the more they circulate, the more affective they become, and the more they *appear* to 'contain' affect."[56] The term itself does not possess this affective energy, but rather, according to Ahmed, it acts as a node in the relational attachment of emotion. "Emotions are relational: they involve (re)actions of relations of 'towardness' and 'awayness' in relation to such objects," she writes.[57] Discussions of emotion and affect, then, are always discussions of orientation and the relationships catalyzed, cultivated, or cleaved apart in the process. To analyze the emotional dynamism of a term, Ahmed proposes a dual-axis methodology. The first axis refers to the attachment of emotions to figures—they *stick* and *slide* sideways between signs and objects. What sticks, though, is affected by what has stuck in "longer histories of articulation."[58] Emotion slides sideways between figures, "as well as backwards, by reopening past associations," and so what sticks "is bound up with the 'absent presence' of historicity."[59]

Ahmed's methodology clearly resonates with McGee's dyad of diachronic and synchronic. He writes, for example, that "[diachronic] meanings give [ideographs] some stability, in its historical precedents, while [synchronic] meanings give it *vitality*."[60] The terms McGee uses to describe the synchronic side—*vitality, force*—offer compelling reasons to conceptualize the ideograph along affective and emotional lines. McGee, after all, encouraged critics to situate rhetoric phenomenologically, and to theorize its materiality not as a distinct object, but rather "as a palpable and undeniable social and political *force*."[61] With the ideograph, "attention is called to the *social*, rather than rational or ethical, functions of a particular vocabulary."[62] In other words, ideographs are affective catalysts and guides for emotional orientation that help condition belief and behavior, simultaneously constituting and reflecting social and political norms.

It's been shown how in both its history of articulation and etymological formation, *Anthropocene* positions the human species as culpable for ecological upheaval. This contextual work indexes some of the *Anthropocene*'s rhetorical associations over time, or what McGee would characterize as its diachronic bearing. In delineating the trajectories prompted by guilt and shame, we see the importance of whether climate change is attributed to the species rather than to particular acts or offending groups. The "towardness" and "awayness" of guilt and shame, to adopt Ahmed's phrasing, can be markedly different. In applying an ideographic lens to *Anthropocene*, our attention is called to the term's affective, socializing force. How does *Anthropocene* motivate group identification and to what end? In calling for us to take responsibility for ecocide, what emotional orientations does the term invite? Although the overview of how guilt and shame operate rhetorically provides some answers these questions, the majority of analyses referenced tend to focus on an individual body feeling shame or guilt. *Anthropocene*, however, operates ideographically, conditioning *collective* belief and behavior and emotional dynamics change as they scale, as is evidenced by the unique character of collective guilt. In the final section I draw again on Sara Ahmed, and further examine the rhetorical implications of *Anthropocene*'s orientation toward shame.

A Collective of Well-Meaning Individuals

When the term *Anthropocene* is used to establish a culpatory link between collective human activities and environmental harm, it not only locates the human species as responsible agent, it also shapes the category of "human species" as a rhetorical construct. The term's circulation—and the affective intensity it accrues through that circulation, as well as the narratives that become "stuck" to it—invites one to identify as a member of the human species. Just as nationalist discourses bring "the nation" into being through the bonds of identification, so too with the group, "human species." Identifying with the *anthropos* in *Anthropocene* is a rhetorical decision, even though the biological connotations of *species* might suggest that the groups are an ontological given, resistant to rhetorical configuration. We are firmly in the rhetorical realm when engaging the term.

Ahmed's analysis of how national shame functions is an instructive analogue to consider, worth quoting in full:

> National shame can be a mechanism for reconciliation as self-reconciliation, in which the 'wrong' that is committed provides the grounds for claiming a national identity, for restoring a pride that is threatened in the moment of recognition, and then regained in the capacity to bear witness. Those who witness the past injustice through feeling 'national shame' are aligned with each other as 'well-meaning individuals'; if you feel shame, you are 'in' the nation, a nation that means well. Shame 'makes' the nation in the witnessing of past injustice, a witnessing that involves feeling shame, as it exposes the failure of the nation to live up to its ideals.[63]

In this briskly sequenced articulation, Ahmed illuminates a process that is playing out with the *Anthropocene*'s ideographic circulation. In mapping her analysis onto *Anthropocene*, we see that the term exhorts you to bear witness to ecological devastation. It provides an opportunity for you to take a moral stance on that devastation through your emotional orientation: feeling badly is a way of affirming ecocide as morally wrong. Crucially, however, the invitation's premise is that you denounce ecocide as a representative of the human species, the collective deemed culpable for the harm. Feeling badly is an affective acknowledgment that such destruction is wrong, which allows you to claim an environmental identity as someone who is morally against harming the earth. In ascribing under these terms, however, *Anthropocene* frames your culpability as an intrinsic feature of the species, the result of who you are—***human***.

Because you can't change your species, those bad feelings tend to get coded as shame, rather than guilt. Ahmed's analysis reveals, however, that in the emotional ecology of collective shame, there is an opportunity to regain what was lost through a feeling of collective pride. To paraphrase Ahmed, if you feel shame about causing global species extinction, you are a part of the geologic force known as the human species. *Anthropocene* constructs the human species and by identifying as a member, allows you to *feel* the species' earth-changing powers. Feeling shame in the Anthropocene brings you into alignment with others who feel that ecocide is wrong, prompting a sense of belonging—even a subtle sense of pride for being part of a group that is well-intentioned, but also tragically confined to a fate they can't change.

Notes

1. Paul Crutzen, "Geology of Mankind," *Nature* 415 (January 3, 2002): 23. Biologist Eugene Stoermer coined the term in the 1980s, though it didn't receive much notice. The notion of renaming the epoch to reflect the effects of human activity predates Stoermer, however, as Antonio Stoppani proposed the "Anthropozoic era" in the nineteenth century.
2. Will Steffen, et al., "The Anthropocene: Conceptual and Historical Perspectives," *Philosophical Transactions of the Royal Society* 369 (1938, 2011): 842.
3. Christophe Bonneuil and Jean-Baptiste Fressoz, "Welcome to the Anthropocene," *The Shock of the Anthropocene*, trans. David Fernbach (London: Verso Books, 2017): 5.
4. Jonathan Foley, "The Age of the Anthropocene: Should We Worry?" *New York Times*, May 23, 2011, https://www.nytimes.com/roomfordebate/2011/05/19/the-age-of-anthropocene-should-we-worry.
5. Edward Burtynsky, *The Anthropocene Project*, https://www.edwardburtynsky.com/projects/the-anthropocene-project/.
6. Bruno Latour, "Agency at the Time of the Anthropocene," *New Literary History* 45 (2014): 1–18.
7. *Anthropocene Feminism*, ed. Richard Grusin (Saint Paul: University of Minnesota Press, 2017).
8. Amy Propen, *Visualizing Posthuman Conservation in the Age of the Anthropocene* (Columbus: Ohio State University Press, 2018).
9. Jedediah Purdy, *After Nature: A Politics for the Anthropocene* (Cambridge: Harvard University Press, 2015): 2.
10. Ibid., 2. Emphasis added.
11. Ibid., 4.
12. Michael Calvin McGee, "The 'Ideograph': A Link Between Rhetoric and Ideology," *Quarterly Journal of Speech* 66 (1980): 5.
13. Timothy Morton, *Dark Ecology: For a Logic of Future Coexistence* (New York: Columbia University Press, 2014): 12–13.
14. Ibid., 14.
15. Ibid., 18.
16. Donna Haraway, *Staying with the Trouble: Making Kin in the Chthulucene* (Durham: Duke University Press, 2016): 49.
17. Ibid., 49.
18. Ibid., 47.
19. Ibid., 47.
20. Ibid., 49.
21. Ibid., 55

22. See *Anthropocene or Capitalocene? Nature, History, and the Crisis of Capitalism*, ed. Jason W. Moore (Oakland: Kairos/PM Press, 2016); Bruno Latour, "War and Peace in an Age of Ecological Conflicts," Lecture, Peter Wall Institute, University of British Columbia, September 23, http://www.brunolatour.fr/node/527.html/.
23. Derrick Jensen, *The Myth of Human Supremacy* (New York: Seven Stories Press, 2016): 270.
24. Ibid.
25. Ibid. Jensen defines civilization "as a culture—that is, a complex of stories, institutions, and artifacts—that both leads to and emerges from the growth of cities (*civilization*, see *civil*: from *civis*, meaning *citizen*, from Latin civitas, meaning city-state), with cities being defined—so as to distinguish them from camps, villages, and so on—as people living more or less permanently in one place in densities high enough to require the routine importation of food and other necessities of life." Because cities are by definition unsustainable, and civilization is defined by the perpetual growth of cities, civilization—understood as only one form of culture, not *all* culture—is inherently unsustainable. *Endgame, Vol I: The Problem of Civilization* (New York: Seven Stories Press, 2006): 17.
26. Ibid., 270.
27. Jeremy Davies, *The Birth of the Anthropocene* (Oakland: University of California Press, 2016): 7.
28. Ibid.
29. Purdy, *After Nature*, 4.
30. Ibid., 3. Emphasis added.
31. Purdy, *After Nature*, 4.
32. Because I maintain that the body's sensorium both learns and produces forms of knowledge, to say that "both shame and guilt are acknowledgements of wrongdoing" includes the notion that your skin may register shame while a cognitive inner-narration may miss or repudiate that interpretation—or vice versa. Sara Ahmed writes, for example, that "[w]hen shamed, one's body seems to burn up with the negation that is perceived (self-negation); and shame impresses upon the skin, as an intense feeling of the subject 'being against itself'." *Cultural Politics*, 103.
33. Donald Nathanson, "A Timetable for Shame," *The Many Faces of Shame*, ed. Donald Nathanson (New York: The Guilford Press, 1987): 4.
34. "Shame," Etymonline.com, accessed March 2, 2018, https://www.etymonline.com/word/shame#etymonline_v_23337.
35. Charles Darwin, *The Expression of Emotions in Man and Animals* (New York: D. Appleton and Company, 1897): 320.
36. Alessandra Fussi argues, for example, "Scholars who characterize Greek culture as a culture of shame point out that, while for us guilt and shame

are two interrelated concepts, the absence of the concept of guilt in classical philosophy and literature ought to play a role in our attempt to understand *aidos/aischyne.*" "Aristotle on Shame," *Ancient Philosophy* 35 (2015): 113. Guilt's evolution as an emotional category was gradual. Jennifer Jacquet notes that "the word for *guilt* does not appear in the Hebrew Old Testament [and that] Shakespeare used the word *guilt* only 33 times, while he used *shame* 344 times." *Is Shame Necessary* (New York: Vintage, Penguin Random House, 2016): 29.

37. Herant Katchadourian, *Guilt: The Bite of Conscience* (Stanford: Stanford University Press, 2010): 135.
38. Morton, *Dark Ecology*, 21.
39. Ibid., 21.
40. Ibid., 39.
41. Ibid., 15.
42. Ibid., 23.
43. Ibid., 24–25.
44. Ibid., 24.
45. Ronda L. Dearing, Jeffrey Stuewig, and June Price Tangney, "On the Importance of Distinguishing Shame from Guilt: Relations to Problematic Alcohol and Drug Use," *Addictive Behaviors* 30, no. 7 (2005): 1394.
46. Ibid.
47. Purdy, *After Nature*, 2, 4. Emphasis added.
48. Michael Calvin McGee, "The 'Ideograph': A Link Between Rhetoric and Ideology," *Quarterly Journal of Speech* 66 (1980): 5.
49. Ibid., 6.
50. See Celeste Condit and John Lucaites, *Crafting Equality: America's Anglo-African Word* (Chicago: University of Chicago Press, 1993); Dana Cloud, "'To Veil the Threat of Terror': Afghan Women and the 'Clash of Civilizations' in the Imagery of U.S. War on Terrorism," *Quarterly Journal of Speech* 90 (2004): 285–306; Hugh Miller, "The Ideographic Individual," *Administrative Theory and Praxis* 26, no. 4 (2004): 569–488.
51. McGee, "Ideograph," 12.
52. Ibid., 10–11.
53. Ibid., 12.
54. Michael Calvin McGee, "A Materialist's Conception of Rhetoric," *Rhetoric, Materiality, and Politics*, eds. Barbara Biesecker and John Lucaites (New York: Peter Lang, 2009): 21.
55. Kevin DeLuca, *Image Politics: The New Rhetoric of Environmental Activism* (New York: The Guilford Press, 1999): 37.
56. Ahmed, "Affective Economies," *Social Text* 79.22, no. 2 (2004): 120.
57. Ahmed, *Cultural Politics*, 8.

58. Ibid., 126.
59. Ibid., 45. The influence of psychoanalytic theory on Ahmed's thought indicates that the movement of backward and forward is linked with memory, repression, and sublimation. What sticks in the present, then, may exist for reasons submerged in cultural memory in ways that remain unknowable to analysis.
60. Michael Calvin McGee, "Ideograph," *Encyclopedia of Rhetoric*, ed. Thomas Sloane (Oxford: Oxford University Press, 2001): 378. Emphasis added.
61. Barbara Biesecker and John Lucaites, "Introduction," *Rhetoric, Materiality, and Politics*, eds. Barbara Biesecker and John Lucaites (New York: Peter Lang, 2009): 3. Emphasis added.
62. McGee, "The 'Ideograph'," 9. Emphasis added.
63. Ahmed, *Cultural Politics*, 109.

CHAPTER 5

Guilty Grief and Ecological Mourning

Abstract This chapter argues that grieving for ecological loss is complicated by feelings of guilt for that loss. Without a framework that accounts for the interweaving of guilt, grief, and mourning, we risk feelings of ecological grief remaining individualized, unresolved, and unarticulated, resulting in a form of melancholia. Analysis of the film *Albatross* by artist Chris Jordan reveals a constructive approach to environmental grief and, by extension, I argue, to guilt as well. In this approach, guilt and grief are framed as teachers, helping orient us toward love, care, and connection with the ecosystems that sustain life.

Keywords Guilt · Grief · Mourning · Environmental melancholia · Emotional ecologies · Solastalgia

"On the mornings that had once throbbed with the dawn chorus of robins, catbirds, doves, jays, wrens, and scored of other bird voices there was now no sound; only silence lay over the fields and woods and marsh," Rachel Carson laments in *Silent Spring*. Published in 1962, *Silent Spring* cultivates a presence of absence, inducing attunement to the sounds that characterize a fecund ecosystem. Carson's investigative study not only established a link between chemical biocides and the silencing of life they left in their wake, it also forged an enduring connection between effective environmental advocacy and the compelling depiction of ecological loss.

T. Jensen, *Ecologies of Guilt in Environmental Rhetorics*,
Palgrave Studies in Media and Environmental Communication,
https://doi.org/10.1007/978-3-030-05651-3_5

The core challenge of conveying such damage remains the same today as it did then: how to communicate such loss and its causes in ways that move people to care, yes, but also to take actions that effectively address the problem. The sheer scale, scope, and speed of ecological loss, however, injects additional urgency to this rhetorical work, complicates recommendations for how to adequately respond, and raises the stakes for emotional entanglement. In the time elapsed since *Silent Spring*'s publication, freshwater fish populations have declined by more than 80%[1]; monarch butterflies have plummeted by 90%[2]; more than 40% of the world's insect species have declined at rates that threaten them with extinction[3]; and in just the last two years, more than half of the Great Barrier Reef has been bleached due to increasingly warm and acidified ocean waters, putting thousands of marine life species that depend on the reef at great risk.[4] From sea ice to salmon, old-growth rainforest to local patches of greenspace converted into apartment complexes, ecological loss takes a wide variety of forms, and as our climate continues to destabilize, both the range and rate of ecological loss will increase. Our affective and emotional responses to such loss, whether experienced directly or through its representation, are a pivotal factor in how—or whether—we act to stop, mitigate, or prevent additional harm.

For those who have established bonds of affection for the landscapes and biotic communities being decimated, feelings of anguish, sorrow, and regret are commonplace. The reverberations of ecological loss are felt particularly by those who are deeply engaged with environmental education, activism, and attunement. "For researchers, scientists, scholars, artists, and activists working in areas related to extinction, climate change, ocean acidification, deforestation, biodiversity loss, desertification, and large-scale resource extraction," Ashlee Cunsolo notes, "the work and the knowledge can be existentially and emotionally overwhelming."[5]

Although the ecological devastation has produced a prevalence of environmentally related grief, "there is a surprising lack of discussion around mourning related to environmental loss or dispossession in broader discursive frameworks or public dialogues."[6] This relative dearth of discussion is directly connected to the paucity of emotional frameworks and resources available for grappling with this increasingly pervasive sense of heartache for our rapidly declining natural world. Not attending to the bodily intensities and persuasive capacities of grief, guilt, and mourning has numerous negative consequences. Those who become acquainted with ecological loss face very real risks, including debilitating depression, cynical resignation, activist burnout, and a grave sense of isolation, as Aldo Leopold suggests when he writes, "One of the penalties of an ecological education is that one lives alone in a world of wounds."[7]

Increasingly, however, the significance of grief in relation to ecological upheaval is being recognized. One recent NBC News headline reads, "'Climate Grief': The Growing Emotional Toll of Climate Change."[8] Similar titles echo across major media outlets, such as "Growing 'Ecological Grief' is the Mental Health Cost of Climate Change"[9] and "Facing Down 'Environmental Grief'."[10] Canadian policy-makers were briefed on the matter in 2018, and informed that "[c]limate-related weather events and environmental change have been linked to elevated rates of depression, anxiety, and pre- and-post-traumatic stress; increased drug and alcohol usage; and increased suicidal ideation, suicide attempts and death by suicide."[11] As a result, the briefing concludes, "mental health considerations are likely to be increasingly included in climate vulnerability and impact assessments."[12] These reports indicate a growing openness to discussing grief in relation to ecological crisis, while also signaling its unfamiliarity through the use of quotation marks, offsetting the phrases "climate grief" and "ecological grief" to suggest a relative lack of established acceptance. Nevertheless, this evidence makes clear that ecological degradation is moving beyond environmentalist frames, and being discussed in terms of public health and governmental policy.

The small but increasing attention to ecological grief in news and policy reports is driven in part by a recent spate of scholarship exploring the topic. As Joshua Trey Barnett notes, "[for] many scholars in the ecological humanities, the question of mourning for non-human animals and for the more-than-human world has become crucial for thinking about earthly coexistence in the 'anthropocene,' which [Timothy] Morton recently re-named the 'age of mass extinction'."[13] The edited collection, *Mourning Nature: Hope at the Heart of Ecological Loss and Grief*, represents some of the most sophisticated and forward-looking inquiries into the affective and emotional forces generated by ecological crisis. Editors Cunsolo and Landman advocate for new theoretical and ethical frameworks for understanding environmental devastation as inextricable from the work of mourning, and call for "productive and positive action through individual and collective expressions of ecological grief."[14] This will not be easy, they add, gesturing to the cascading implications of developing new emotional orientations:

> This ecological work of mourning will, necessarily, look different. And it will require from us different forms of commitment, different attention, and different ways of thinking. It will require us to engage with ourselves, with others, with non-humans, and with emotions differently.[15]

This work will also require engagement with feelings of guilt for the lives being mourned. In fashioning new frameworks calibrated to the realities of mass extinction and climate crisis, it is critical to acknowledge how feelings of complicity for ecological loss will, at the very least, alter the dynamics of mourning. In many cases, however, the combination of guilt and grief can significantly impede the transformative work of mourning, such that one's grief remains individualized, unresolved, and unarticulated.

At stake here is the risk of slipping into a state of environmental melancholia, "wherein one may be 'frozen' or otherwise arrested … in a form of perpetual, unresolved mourning."[16] In *Environmental Melancholia: Psychoanalytic Dimensions of Engagement*, Renee Lertzman describes the condition as one "in which even those who care deeply about the well-being of ecosystems and future generations are paralyzed to translate such concern into action."[17] Attending to the complexities of collective guilt can strengthen the productive work being done around ecological mourning, and enhance our capacity to engage the interwoven emotions of guilt and grief. Not doing so threatens to attenuate the usefulness of new frameworks. Moreover, we risk squandering the political potential in mourning: when painful, disorienting feelings of loss experienced by an individual (grief) are worked through in ways that promote reengagement with the world post-loss (mourning), they can become a potent catalyst for activism and expressions of ecological care.

There are considerable challenges to establishing ecologically attuned forms of grief and mourning, which I outline in the following section. These challenges parallel those faced in creating new models of guilt. This chapter argues, therefore, that the creative and critical work of environmental scholars and activists around grief and mourning offer productive routes for doing the same with guilt. Developing these emotional literacies in tandem helps to further illustrate how emotions work ecologically, both metaphorically and materially. After indexing some of the key challenges involved, I analyze the film *Albatross*, which defies many conventions of the nature documentary genre. This film offers a fresh perspective on environmental grief and, by extension, I argue, to guilt as well. In this approach, guilt and grief are framed as teachers, helping orient us toward love, care, and connection with the ecologies that sustain life.

Challenges of Guilty Grieving

In *Dark Ecology*, Timothy Morton likens ecological attunement to a noir narrative, but with a twist ending. More accurately, it's less of a twist than a strange loop, wherein a detective investigates manifold harms done to the environment, propelled by the connective capacity of ecological thinking, only to uncover their own complicity in the crime. Ecological awareness, Morton asserts, is the moment of realization when one discovers they are both detective ***and*** criminal.[18] When this awareness of culpability accompanies knowledge of ecological loss, as it so often does, that unique sense of guilt becomes entwined with other emotional responses to loss. The conflicted position of being both detective and criminal is mirrored on an emotional level: ecological losses often elicit grief and guilt, summoning urges to mourn ***and*** atone. As Nancy Menning wisely observes, "We must mourn not only what we have lost, but also what we have destroyed."[19]

It is not unusual for guilt to be an element of one's grief. Following the death of a loved one, we are prone to imagine what we would have done differently. We may regret things left unsaid or imagine actions we might have taken that could have prevented or delayed the passing. Guilt can be considered a normal part of grieving, not only because it regularly appears, but also because grief itself is not a singular emotion, but rather an active assemblage of emotional components. Grief is composed of multiple feelings, with fluctuations of anger, sadness, fear, numbness, confusion, regret, and/or disgust, to list some of the more common states. Guilt's role within grief's emotional cluster, however, is unique. A review of psychological studies on grieving found a "[h]igh prevalence of guilt [in] several quantitative studies and its detrimental potential in relationship to psychological and physical health and general well-being has been confirmed by empirical evidence."[20] While guilt's volatility and its potential for complicating the grieving process has been proven, the same study also emphasizes the difficulties of analyzing guilt. "The lack of conceptual convergence on guilt is well-illustrated," in part, they note, because "psychologists have not reached agreement on defining guilt."[21]

Guilt's plasticity allows it to take many forms, and this malleability makes it difficult to predict precisely how it might impact processes of grieving.

The entanglement of grief and guilt—common, but also variable—warrants their co-examination, particularly for ecological loss, since we are often connected to the cause of that loss through our participation in industrial economic systems. The benefits we might receive from these systems, real or imagined, amplify the likelihood of guilt and consequently, create conditions ripe for environmental melancholia. Renee Lertzman explains that "competing affective investments" generate conflicting responses toward ecological loss, which "can fracture or divert psychic energies in terms of adequately processing a loss."[22] Lertzman describes the tangle of contradictory affects and identifications through a psychoanalytic frame:

> grievances experienced as a result of the loss and disappointment (e.g. contaminated water, the pollution of a beach, the muck of algae) are not directed outwardly but rather remain private and contribute to a subjectivity that may be incapable of action or repair. In order to mourn … the subject needs to be able to recognize the loss [but if there is an] unclear origin of the loss and an inability to articulate this socially … it may not be possible to undertake the work of mourning [which constitutes] a form of *melancholic* subjectivity, impeding what might otherwise be seen as viable and active engagement with the issues.[23]

The collective guilt experienced for playing some role in the ecological loss, however small, can create affective and cognitive dissonance, that when entwined with grief, leaves one feeling emotionally stuck. The seemingly irresolvable forces create a sense of isolation, frustrating grief's transformation through mourning. Lertzman keenly notes that this perturbed emotional state is often framed as *non-engagement* with environmental issues, whereas it might more productively be viewed as a state of stalled or obstructed engagement. Guilty grieving is precarious work, after all. Shifting how we frame its frustrated form—from apathy to impeded mourning—is the first step toward developing tools and literacies for helping navigate ways out of it.

Awareness of complicity in ecological loss brings additional complications to a grieving process already facing significant obstacles. Chief among them are Western cultural norms that have, to put it mildly, unevenly distributed what counts as "mournable." Social pedagogies explicitly and implicitly instruct who is and who is not considered worthy of grief, boundaries that are built and enforced through myriad mechanisms. Consider a phrase like *tree-hugger*, for example, which has long been used to disdain environmentalists. The implied message isn't subtle: the tree is an object for which one should not express affection. To diminish affection is to also diminish the possibility of mourning. Even grammar conventions contribute to such boundaries. "Usually, the English language reserves the pronoun *who* for humans and uses *that* for nonhumans," Derrick Jensen notes in *The Myth of Human Supremacy*, observing that this linguistic habit signals "meaningful subject and meaningless object."[24] Robin Wall Kimmerer similarly critiques contemporary English's "grammar of animacy," which rarely extends to plants and animals, let alone whole landscapes. "In English," she writes, "we never refer to a member of our family, or indeed to any person, as *it*. That would be a profound act of disrespect. *It* robs a person of selfhood and kinship, reducing a person to a mere thing....English doesn't give us many tools for incorporating respect for animacy. In English you are either a human or a thing."[25] Through myriad means, our emotional ecologies are shaped—partially, but powerfully—through social pedagogies that invoke rules of what is an acceptable subject of affection and by extension, of mourning.

Expanding the boundaries of mourning to include nonhumans is a key challenge in the emotional work of ecological loss. The intricacies of our imperiled ecologies make some of this work more difficult than others. Including charismatic megafauna such as polar bears and orcas as mournable bodies is one thing. Mourning the plummeting populations of zooplankton in major watersheds, on the other hand, presents a different set of rhetorical barriers. "[A]lthough it is hard to imagine the loss of a human," write Cunsolo and Landman, "it is harder still to think through the loss of non-humans, particularly if the non-humans we are mourning are large assemblages or systems, such as a body of water or an ecosystem or a forest."[26] Expanding the boundaries of mournable bodies will require substantial rhetorical innovation in a variety of forms.

The challenges of creating new modes for navigating and transforming what I've called guilty grief are also opportunities for rhetoricians drawn to an emerging frontier of environmental communication. As shown in the following section, mourning is a profoundly rhetorical process. This area of study is sure to rise in importance, since "[w]e are entering a time when ecologically based mourning seems likely to occupy more and more of our experience."[27] We should read *challenge* here in both of its senses, then: as a call to engage *and* as a demanding trial that tests our character.

Between Grief and Mourning, Rhetoric

In *Understanding Grief*, Richard Gross notes that while grief and mourning are often used interchangeably in everyday discourse, there are important distinctions. Whereas grief refers to an individual's psychobiological reaction to a significant loss, mourning refers to "a longer-term process," in which the bereaved "attempts to redefine the relationship to the deceased, their sense of self, and the external world."[28] Gross adds that "mourning is also used in a very different sense" to "denote the culturally defined ways of publicly expressing one's grief, as in rituals and customs related to the funeral."[29] Mourning, in other words, is a process of reorienting to a world without the loved one, which often entails conveying grief in collaboration with others through a diverse set of cultural practices that aim to positively moderate and transform grief. These practices are acutely rhetorical in design and implementation, making mourning a deeply rhetorical process.

Unsurprisingly, the rhetorical tradition is dotted with studies of mourning's rhetorical forms. In *The Emotions of the Ancient Greeks*, David Konstan observes that the Greeks separated grief from mourning, and had multiple terms for different shades of their experience, including *lupê*, translated as distress, and *penthos*, which referred to the public expression of grief, often by way of song.[30] He takes note of the "many forms" of mourning practices, "from sobs and ululation to ritualized actions such as the tearing out of hair and the beating of one's breast," adding that "[an] entire genre, the so-called consolation, was developed to help people recover from sorrow over the loss of a loved one," further demonstrating the depth and detail of rhetorically crafted ritual.[31]

Invoking the historical lineage of rhetoric's involvement with the artful transformation of grief, Beth Hewett argues that the field needs to reengage mourning rituals as part of its civic portfolio. The eulogy, one of the most common mourning practices, etymologically means "good words," yet Hewett notes that in grief, "people naturally feel torn apart [and often] feel verbally inarticulate."[32] Such "ceremonial occasions represent an area in which rhetoricians can use rhetorical knowledge and abilities to assist [others] in times of grief and mourning"—and moreover, if the opportunity arises, they ***should***, as "contemporary rhetoricians [have a] responsibility for public outreach in civic and educational forums."[33] Recalling Cunsolo and Landman's comment that ecological mourning demands "different forms of commitment, different attention, and different ways of thinking," I echo Hewett's call and suggest that rhetoric studies is well-positioned to help create new modes of mourning.[34]

The rhetorical practices of mourning have been developed to provide guidance and structure for grief, with the aim of transforming grief's affective intensity in reorienting after great loss. Konstan and Hewett show rhetoric's connection to mourning at historical polarities, spanning ancient Greece to the twenty-first century. They also gesture to reasons for why rhetorical transformations of grief were—and remain—so important. Grief's intensity can be volatile, particularly when entwined with guilt. As Donovan Och notes in *Consolatory Rhetoric*, "[if] the psychological stress caused by the experience of grief, personal and collective, is left unattended," she warns, "consequences of the most dire sort" can happen: Suicide, psychotic response, severe depression, loss of health, or a state of total helplessness do occur if grief is unresolved.[35] Grief that is blocked from healthy mourning practices can be unpredictable and unsafe. Such is our current moment. We are in the utterly surreal position of learning how to mourn the loss of entire species.

Evidence that ecological grief is already exacting a toll has escalated in the past few years. The American Psychological Association, for example, has been mobilizing resources around "ecoanxiety," releasing a 70-page report in 2017 titled "Mental Health and Our Changing Climate." The report's opening statement provides a stark frame of reference for what's at stake:

> The health, economic, political, and environmental implications of climate change affect all of us....They induce stress, depression, and anxiety; strain social and community relationships; and have been linked to increases in aggression, violence, and crime....[The] psychological responses to climate change, such as conflict avoidance, fatalism, fear, helplessness, and resignation are growing. These responses are keeping us, and our nation, from properly addressing the core causes of and solutions for our changing climate.[36]

The APA report suggests there are three prominent patterns of response to awareness of pressing ecological issues: acceptance leading to debilitating despair, acknowledgement via avoidance, and acceptance leading to active address of core causes, supplemented with emotional and psychological support mechanisms. These response patterns obviously overlap with one another, and often in the same person.

One poignant example of the intensity and complexity of emotional response to ecological crisis is found in the suicide of David Buckel, who in April of 2018, set himself on fire in a park in New York City. Buckel, a committed environmental activist and prominent civil rights attorney, hoped that his death would act as a mirror, writing in his final message, "my early death by fossil fuel reflects what we are doing to ourselves."[37] Buckel's letter indicates that a key source of his despair, anger, and grief, was the lack of action in others *despite* their awareness of the problem. "Many who drive their own lives to help others often realize that they do not change what causes the need for their help," he writes.[38] Buckel characterizes his self-immolation as a result of seeing "no other action [that] can most meaningfully address the harm [I] see."[39] As noted in the opening chapter, collective guilt is often experienced as feeling like you are not doing enough to help. Dominant conceptions of guilt often focus on doing wrong, rather than failing to do right, whereas feelings environmental guilt frequently take the latter form.

Buckel's story also tragically illustrates an interwreathing of emotional ecologies. He aimed to have the horror of his self-immolation shock others into being horrified at the escalating ecological death. His actions were motivated by his love for the Earth and its inhabitants, deep grief for their systematic destruction, disappointment and anger at others for not acting on their knowledge of environmental crisis, and guilt over not being able to do more himself. An ecological perspective encourages attunement to how emotions interconnect, particularly where emotions take other emotions as their object (anger about apathy) and the intensities of one feeling fueling another (grief feeding guilt). In advocating for an ecological perspective in the rhetorical study of emotion, I am trying to push our analyses to embrace the writhing tangles of emotions working with and against each other in dynamic interactions—not so that we might untangle and carefully catalogue the parts, but rather, to see systems instead of singularities. The stakes for doing so are not inconsequential. As Buckel's suicide and the APA report demonstrate, our capacity to navigate environmental grief and guilt can be, at their extreme, a matter of life and death.

Environmental activists, artists, and scholars engaged with grief offer several constructive routes for making new modes of ecologically attuned mourning accessible. In the following sections I highlight two particularly productive models for developing emotional literacies aligned with the ecological realities of our age: Glenn Albrecht's efforts to expand terminology around *earth emotions* and Chris Jordan's exploration of grief in the film *Albatross*.

A Lexicon of Ecological Mourning

The boundaries of mourning, as previously noted, are created and reinforced in part through everyday language. Seemingly mundane semantics help fashion social pedagogies that, to date, have reserved mourning practices for the loss of other humans. Changing linguistic norms is therefore a vital component in creating new modes of ecologically attuned living. Robin Wall Kimmerer and Derrick Jensen illustrate how part of this work involves identifying anthropocentric conventions and altering them to engender more just and inclusive modes of relation. Enriching our ecological lexicon is also essential, a process that entails both recouping nomenclature and fashioning new terminology that identifies and elucidates ecological elements, exchanges, and emotions. Glenn Albrecht has done substantial work in the latter category, developing new terms to explain psychological and somatic relationships we have with the earth. The creation of this corpus was catalyzed by ecological loss, and the linguistic limitations Albrecht encountered in trying to articulate it.

Albrecht's research with communities affected by environmental calamities—towns located next to mountain-top removal mining and ranching homesteads hit with climate change-driven drought—brought him in contact with people experiencing a "place-based distress," which they struggled to describe.[40] In observing closely, Albrecht saw they had "a similar melancholia as that caused by genuine nostalgia," a kind of doleful longing to return home, yet they were still living in their home. To describe the emotional experience, Albrecht coined the term *solastalgia*, defined as "the pain or sickness caused by the ongoing loss of solace and the sense of desolation connected to the present state of one's home and territory."[41] In other words, solastalgia is the "homesickness you have when you are still located within your home environment"—but a home environment that has been marred by ecological disorder.[42]

The neologism quickly spread, and has since been used in news reports, scholarship, policy documents, and creative art projects across the globe. The rapid uptake of the term indicates that it captured a pervasive feeling experienced across geography and culture, a collective "grieving for the loss of the signs and symbols of a healthy ecosystem and place."[43] Solastalgia gives voice to a public feeling generated by ecological disruption, empowering many to articulate their feelings of disempowerment, to recognize how many others share the sentiment, and to collaboratively deliberate on how to respond.

"Despite the importance of connections between environmental health and human health (physical and mental) in many cultures, we have very few concepts in English that address environmentally induced mental distress," Albrecht writes.[44] Other cultures, especially indigenous peoples with long-established relationships to the land that supports them, have "concepts in their language that relate psychological states to states of the environment."[45] An impoverished emotional vocabulary constrains our capacity to observe, appreciate, and understand how our feelings orient us, shaping our relations with others, human and more-than-human. Inability to articulate emotional states, particularly those that seem more novel or subtle, hinders connection with others, even when they might feel akin. Albrecht's lexicon of ecological emotions exemplifies a productive pathway toward emotion literacies that more holistically connect our health with the health of our environment.

Situated within this typology is *new mourning*, which, in contrast to his inventory of neologisms, is characterized more as an extension and amplification of existing mourning structures. Albrecht argues that the experience of grief has changed in profound ways and as a result, mourning practices must change with them or risk the negative outcomes of grief turned paralytic or worse. What distinguishes this new, ecologically attuned mourning from the dominant modes of mourning in Western traditions? Albrecht identifies four elements, one of which— "enhanced empathy for the non-human"—I discussed earlier in outlining challenges to expanding mourning practices.[46] The other three elements that characterize new mourning include "detailed knowledge of causality, anthropogenic culpability [and] feelings of powerlessness."[47] These three features all share a common denominator: environmental guilt.

The first factor highlights knowledge that links loss to causal agents, a form of knowledge, I've argued, that can expand rapidly as one cultivates ecological attunement. The second element, "anthropogenic culpability," indicates a sense of complicity that includes both cognitive and affective awareness of one's participation in massively distributed acts of violence. And the third, powerlessness, highlights a commonplace feeling that accompanies such a capacious, collective guilt. Albrecht traces this helplessness to an incommensurate relationship between negative feelings over ecological loss and one's perceived ability to respond adequately:

> Powerlessness in the face of pervasive change agents such as multinational corporations and authoritarian governments is another factor in the new mourning. Within the grip of such concentrated centres of power, individuals cannot direct their grief about negative environmental change towards anything or anyone in particular. Such powerlessness (personal and political) is also a defining feature of the concept of solastalgia.[48]

Feelings of powerlessness arise, in other words, when one's sense of guilty grief and desire for making amends are weighed against perceived options for remedying the wrong—and found to be wholly disproportionate. Since the size of the offense one feels guilt and grief for grossly outweighs what one person feels capable of accomplishing, helplessness ensues. As Lertzman argues, this sense of helplessness is in turn frequently deemed "public apathy" and is perceived as an "absence of care towards environmental quality or protection."[49] In contrast to a lack of emotion (*a-pathos*), non-engagement may in many instances be better read as a "psychic defense" to a surplus of complicated emotions, resulting in environmental melancholia.[50]

That our ecological grief is so often a guilty grief is increasingly recognized across emerging work on the issue. In "Climate Change as the Work of Mourning," Ashlee Wilcox claims that the "tension between mourning what has been lost or what is changing coupled with guilt over our own actions that have led to these changes is an important area for further research and consideration."[51] Similarly, Nancy Menning argues that "[e]cological losses differ in important ways from human deaths. In particular, we are often complicit in these losses, if only by virtue of living in the Anthropocene."[52] And as Albrecht makes evident, "anthropogenic culpability" is an integral element in the formulation of "new mourning." These scholars rightly emphasize the meaningful interconnections between grief and guilt in response to ecological loss, yet the dearth of extant critical inquiry into environmental guilt limits the theoretical range of their insights, which means that few are able to go beyond acknowledgement of its importance.

For new modes of ecological mourning to flourish, we must engage environmental guilt's *collective* dimensions. If feelings of culpability remain coded solely as individual guilt rather than collective guilt, the grief we feel over ecological loss is even more likely to remain individualized. And as Lertzman reminds us, when our grievances over contaminated water or clear-cut old-growth "are not directed outwardly but remain private [they can] contribute to a subjectivity that may be incapable of action or repair," which inclines grief toward "inwardly directed impoverishment, unresolved and static."[53] The transformation of grief through rhetorical practices of mourning is in many ways a transition in orientation from individual to the collective. To advance new modes of mourning, then, I argue we must evolve our understanding of environmental guilt, moving beyond heavily individualized conceptualizations and toward a more collective, ecological notion.

Glenn Albrecht's ecological lexicon embodies the kind of emotional literacy development that is critically needed. Despite his focus on culpability and its capacity to modify emotions related to ecological crisis, however, there are currently no terms explicitly conceived to elucidate the dynamics of environmental guilt in his collection, *Earth Emotions: New Words for a New World*.[54] Expanding our vocabulary for discussing the intricacies of environmental guilt is a productive, though mostly untrodden, route forward.

Cultivating the language of emotional ecologies is part of a larger movement toward enriching our ecological vernacular more generally. An important part of this work is recuperative, as our linguistic capacity for engaging ecology is being attenuated alongside our emotional connections for ecological care. Robert Macfarlane provides a prime example of what this labor could look like. In 2007, Macfarlane was made aware of "a culling of words concerning nature" in the *Oxford Junior Dictionary*.[55] "Under pressure, Oxford University Press revealed a list of the entries it no longer felt to be relevant to a modern-day childhood," he writes.[56] Among the list of deletions were *acorn*, *dandelion*, *fern*, *heron*, *ivy*, *kingfisher*, *nectar*, *newt*, *otter*, *pasture*, and *willow*. Among the list of words added: *blog*, *bullet-point*, *chatroom*, *committee*, and *voice-mail*. "A basic literacy of landscape is falling away up and down the ages," Macfarlane laments, and our faculties for ecological observation and appreciation are vitiated in the process, along with "the power that certain terms possess to enchant our relations with nature and place."[57]

In *Landmarks*, Macfarlane provides a model response to this decline, offering a collection of chapters exploring intimate relationships between land, language, and people. From a "Peat Glossary" compiled by Scots to capture the many moods of the moors to neologisms nuancing the acoustic dalliances of bird sounds, Macfarlane furnishes an invigorating account of language's capacity to both freshen and fortify our ecological relationships. Cultivating a lexicon for emotional ecologies that has more depth and nuance is not the end goal, but a means for reorienting—and rejuvenating—our relationship with and within the systems that sustain life itself.

Grief Is the Felt Expression of Love

When artist Chris Jordan stood over albatrosses who had perished from plastics ingestion, he consistently struggled to get the photograph he wanted—his glasses were too full of tears. Following his first visit to Midway island to photograph these deaths, Jordan fell into depression.[58] Around the same time, his photographs of albatross bodies surreally embracing piles of plastic began to circulate widely. As he continued to promote his work, the pain he carried from Midway eventually matured to a point that it threatened to break him entirely. "I've realized that the metaphor that I carried with me when I went to Midway was that Dante's *Inferno* scenario," he recalled, adding, "You walk through the fire, and you come out the other side with renewed energy or perspective. What actually happened, though, is that I felt like I walked through the fire and then just burned up in it."[59] The experience pushed him into an ultimatum: stay locked in depression or discover another way to emotionally engage with the realities of what he witnessed.

In my analysis of Jordan's *Midway* photographs in chapter two, I focus primarily on their composition and the rhetorical situation they construct for the viewer. I argue that the images function as a form of visual enthymeme, providing two premises—the decaying remains of an albatross and a pile of plastics exposed from its stomach—that create a juncture at which viewers are compelled to complete the sequential chain of argument. To conclude this chapter on guilt, grief, and mourning, I consider the artist's follow-up project, *Albatross*, a full-length film that situates the albatrosses and the artist in a wider context. These entwined readings align with the driving impetus behind Jordan's paired projects. The first primarily sought out how to communicate the global plastics problem in an affectively compelling way. *Albatross* also seeks to accomplish that feat, but it expands to contemplate more directly the strong feelings elicited upon witnessing such a visceral scene of ecological loss. Jordan's personal revelation about grief offers one possible answer to the quandary posed by Cunsolo and Landman in "To Mourn Beyond the Human":

> We are entering a time when ecologically-based mourning seems likely to occupy more and more of our experience. And yet it seems that we are also entering a time of great denial and avoidance of this type of work because of what it means and what it will entail *if we were to truly embrace and open ourselves to these changes and this mourning, and to the understanding of our individual and collective responsibility*. We are all implicated in this loss; indeed, our very lives and existence are predicated on the deaths of other bodies that have come before us—human and more-than-human—and on the promise of future deaths. *How do we even begin to think and feel that?*[60]

Albatross opens with the crisp ring of meditation chimes. As the peal ripples in resonant waves that slowly fade away, an ominous tone gathers, replacing silence with a haunting, low-pitched warble that contrasts with the high, clean tonality preceding it. The foreboding groan swells in intensity as the words of Coleridge appear: "He loved the bird, that loved the man / Who shot him with his bow."[61] These lines from *The Rime of the Ancient Mariner* glow in bright white against a black backdrop as the sound crescendos until image and sound are severed, full stop, by the knell of a baritone bell. Slowly, a celestial mandala comes into view while the bell's reverberations ebb and flow.

Albatross signals in its first minute that it will be something other than a traditional nature documentary. The opening moves indicate that its inquiry will be performed as more of a meditation—an extended contemplation of a topic—and through its use of chimes and bells and mandalas as an introduction, it also suggests that viewers enter into a meditative state of their own. The rhythm of the film helps create this state, one that is palpably distinct from other nature documentaries, which tend to rapidly switch camera angles from one perspective to another, jump-cutting between different shots, depicting dramatized scenes of action. In *Albatross*, the camera does not linger, so much as it sits and stays, replicating a meditative pose. Jordan noted that this editing style evolved during the second trip to the island in a way that mirrors aspects of his emotional transformations. "I wanted to convey the intensely vivid, sensual, emotional, and spiritual experience of being with [the albatrosses] on the island," he notes.[62] The film evokes an oceanic, meditative space, layering shots over one another, like waves washing over themselves again and again.

Contemporary nature documentaries, like those filmed for *National Geographic* or PBS, often feature frequent interjections by a voice that guides you through different scenes. In contrast, *Albatross* is sparsely narrated. It is, however, full of sound, with noisy albatrosses and whipping pacific winds. The initial lines of narration do not arrive until several minutes in. Laid over still frames of photographs from Jordan's earlier *Midway* series, we hear the artist's voice slowly rolling out a narrative with pregnant pauses between each sentence:

> Kneeling over these scenes is like looking into a mirror. Here we face one surreal consequence of our collective choice. This is our culture turned inside out. I was drawn to Midway because it offered a chance to face a global issue on a personal scale. I knew it wouldn't be easy. I came here as a witness, with no story, or answers in mind.[63]

Jordan's voice-over implies that viewers adopt a similarly open stance. As the film visually draws viewers into the island through stunning cinematography, the narration suggests one *feel* their way through the journey on a personal scale. As the introduction moves to completion, the voice-over offers a tacit welcome to enter as witness without prejudgment. And with that prompt, viewers are brought into the lifecycle of the albatross, learn its habits, marvel at the bird's endurance and devotion to its partner and chicks, and observe the travails albatrosses face, both natural and industrially introduced. *Albatross*'s evocative soundscapes, sensual cinematography, and sparse but deliberate narration ground the viewer's affective experience as the foundational evidence for deliberating ecological loss.

As the film progresses, viewers learn of the impact of plastics on albatrosses. We are shown a mother regurgitating food into a hungry chick's mouth and when the camera zooms into show the chunks of plastic being unwittingly delivered to the newborn, it becomes quite clear what Jordan meant when he said, "I knew it wouldn't be easy." Though the film never lingers unnecessarily on such distressing scenes, it does not shy away too quickly from them either. In the penultimate scenes, the camera stays affixed on an albatross and its heaving chest, beak languidly swiping the sand, unable to move any other muscle. Another is shown struggling at sea, too weak to fly, but with enough energy to resist the inevitability of drowning a bit longer. These agonizing scenes arrive at nearly an hour and a half into the story, at a point when the identification between viewer and albatross has been given a chance to mature and deepen. Intimately watching any living creature in its last throes is painful and difficult. If the film has successfully brought you into feeling deeply for the albatross by that point, however, the grief can be absolutely piercing.

That feeling, however, is precisely what the film encourages you to sit with and allow to happen. The final lines of narration hone in on the emotional ecologies at work. Over images of deceased birds' limp bodies breezed by shore winds and shrouded in shifting sands, Jordan closes the film with his hard-earned insights on grief:

> The most difficult thing to bear, for me, was what I knew, but they couldn't know, about why they were dying. In this experience, the true nature of grief revealed itself. I saw that grief is not the same as sadness or despair. Grief is the same as love. Grief is a felt experience of love for something we are losing or have lost. When we surrender to grief, it carries us home, to our deepest connection with life.

These final lines invite us to go deeper into our grief, to give ourselves over to the searing agony of ecological loss. In doing so, Jordan avows, we will recognize and be drawn closer to what we love, and our grief will be transformed—not necessarily expunged, but transmuted into deep ecological kinship. *Albatross* offers a provocative model for ecological mourning, wherein grief's affective intensities are reframed as the register of ecological love.

The film's conclusion does not turn explicitly to hope, as is the prevailing convention in environmental communication focused on declining and threatened populations. This is not for want of reflection on the matter. Jordan has articulated a skepticism toward hope's efficacy in multiple interviews, noting, for example, "we have this cultural obsession with hope. I'm not sure how useful hope really is."[64] Appeals to hope in such conclusions tend to function as a form of rescue, ushering viewers to move from feelings of guilty grief to more "optimistic" ground. If audiences exit the story on a dour note, the rationale goes, it is improbable that they will rally for the cause; rather, they will sink into a sulk of despair and helplessness. Hope is often offered as an antidote to grief, relief from the acute pain experienced from witnessing ecological loss. Hope is framed as both salve and spark. Conclusions that operate on this logic suggest that the discomforts of loss can be avoided, *if only we would act.* Moreover, they convey that *it is possible* that these actions will succeed, and *that* is why one should feel motivated.

Albatross does not ask viewers to move away from their grief. To the contrary, its invitation is to *entrust* ourselves to grief's intensities. The acute discomfort of ecological loss is borne out of deep empathetic connection. If that agony is framed as something to be avoided or tranquilized through hope, those ecological connections are removed from the center of attention. By framing grief as the felt expression of love, those painful but powerful feelings are approached as orienting forces, directing us *toward* ecological connection. Rather than proferring hope as the preeminent emotion for relating to loss, *Albatross* places love of ecological relation at the center.

In its reconfiguration of grief and its animating potential, *Albatross* has constructive implications for new modes of ecological mourning. As explained earlier in the chapter, grief's transformation through mourning relies in part on a shift from individual to collective through expressions about the loss. This work is often facilitated through mourning practices and genres rhetorically shaped to accommodate the loss of other humans. As Lertzman notes, ecological grief frequently remains "[i]nchoate because such losses exceed the language needed to express them and mute because of the lack of socially sanctioned forms for sharing emotional responses about ecological issues."[65] Developing language for articulating ecological grief is one productive pathway forward, as is expanding the boundaries of what counts as mournable, and creating new formats and rituals for acknowledging these losses and the grief they generate.

Without this work, we continue to risk grief turning toxic. "Mourning is a social process, and in the absence of sharing it, we remain stuck and our mourning in stasis," resulting in environmental melancholia—"a form of perpetual, unresolved mourning."[66] What *Albatross* suggests, however, is that mourning is not only a social process, but also an *ecological process.* Healthy mourning is most often conceived of as reorientation to the *social* world, in which one's connections with other humans are restored following the loss of a loved one. The film posits a model of mourning where reorientation is to the *ecological relations* that comprise our world, in which one's connections with nonhumans are restored following loss. And when read closely, it is evident that the restorative work goes further: one's *capacity to connect* is strengthened as well. Grief's transformation is still a shift from individual to collective, yet that collective is no longer constrained to the social spheres of humans—it becomes ecological, teeming with more-than-human kin.

Guilt Is the Felt Expression of Care

Jordan's reconfiguration of grief resonates with the theoretical frame of emotional ecologies because it orients us to the dynamic interaction between emotions, engages their phenomenological register, *and* does so in the context of relating more fully with nonhumans. We may grieve the albatrosses whose deaths are driven by plastic, but by releasing our *resistance* to grief, it can orient us to our *ecological connections* with albatrosses and by natural extension, the wider ecologies in which we are enmeshed. And in being oriented to our ecological entanglements, this grieving process works to enrich and strengthen those relations, revealing grief as the felt expression of love for them.

Although guilt is not named explicitly, it is prominent in Jordan's reconfiguration of grief. After nearly thirty minutes without any voice-over narration, the last fifteen minutes of which are artful, agonizing displays of dying and dead albatrosses, we are rejoined with Jordan, whose lead sentence is a complex articulation of collective guilt: "The most difficult thing to bear, for me, was what I knew, but they couldn't know, about why they were dying." Jordan highlights knowledge of causality for ecological devastation and intimates a sense of culpability, which, when combined, generate feelings of environmental guilt. That he rates this awareness and its accompanying affects as the *most difficult thing to bear* is a testament to the power of collective guilt and its interwreathing with ecological grief. It is instructive, then, that upon engaging more deeply, this collective guilt becomes *the source of revelations* regarding grief. "In this experience," he states, of bearing the extraordinary torment of being fully receptive to the cause of the albatrosses' suffering, "the true nature of grief revealed itself." It is through attunement to environmental guilt, in other words, that Jordan reorients his understanding of grief. And the maneuver by which grief is reconfigured at the end of *Albatross* can similarly be applied, I argue, to that sense of guilt.

Just as grief is a felt experience of love for something we are losing or have lost, *guilt is the felt experience of care for something we have harmed.* From this perspective, grief and guilt orient us to what we love and what we care for and about. They become teachers, guiding us toward our affective attachments and emotional investments. Environmental guilt, I'm suggesting, can guide us toward a deeper understanding of our care by illuminating the boundaries that have been transgressed, and by leading us to recognize the ethical systems that create and constellate those boundaries.

When we feel guilty for ecological harm, we are engaged with an ethics that precedes and exceeds cultural systems, yet that is nevertheless influenced by them. The poignant pain we feel in response to an albatross's death by means of plastic, or to the obliteration of a forest to be replaced by human housing, or to the loss of salmon blocked from ancestral spawning grounds by hydropower dams is the expression of ecological connection, as well as cultural interpretation. Just as understandings of grief and mourning need to be revised to acknowledge that our emotional health is inextricably tied to the health of our biospheres, so too must our understandings of guilt.

I've argued that this work is best done in tandem because ecological grief and guilt are profoundly intertwined. The creative and critical work that is occurring around grief and mourning in scholarly and activist communities, moreover, suggest possible models for reconfiguring our notions of guilt within an ecological context. Our sense of guilt for environmental devastation can be a powerful source for revealing and reconnecting with what we care for. As a keystone emotion for contemporary environmental rhetorics, this reconfiguration can have powerful, cascading implications for how we engage ecological crisis.

Notes

1. WWF International, *Living Planet Report 2016: Risk and Resilience in a New Era* (Gland: WWF International, 2016): 30.
2. Wayne E. Thogmartin, Ruscena Wiederholt, Karen Oberhauser, Ryan G. Drum, Jay E. Diffendorfer, Sonia Altizer, Orley R. Taylor, John Pleasants, Darius Semmens, Brice Semmens, Richard Erickson, Kaitlin Libby, and Laura Lopez-Hoffman, "Monarch Butterfly Population

Decline in North America: Identifying the Threatening Processes," *Royal Society of Open Science* 4, no. 9 (2017): n.p.

3. Francisco Sánchez-Bayo and Kris A. G. Wyckhuys, "Worldwide Decline of the Entomofauna: A Review of its Drivers," *Biological Conservation* 232 (2019): 8–27.
4. David Bellwood and Jennie Mallela, "Biodiversity of the Great Barrier Reef," *Australian Academy of Sciences*, www.science.org.au/curious/great-barrier-reef.
5. Ashlee Cunsolo, "Preface: She Was Bereft," *Mourning Nature: Hope at the Heart of Ecological Loss and Grief*, eds. Ashlee Cunsolo and Karen Landman (Montreal: McGill-Queen's University Press, 2017): xvi.
6. Ashlee Cunsolo and Karen Landman, "To Mourn Beyond the Human," *Mourning Nature: Hope at the Heart of Ecological Loss and Grief*, eds. Ashlee Cunsolo and Karen Landman (Montreal: McGill-Queen's University Press, 2017): 5.
7. Aldo Leopold, *Round River: From the Journals of Aldo Leopold* (Oxford: Oxford University Press, 1953): 163.
8. Avichai Scher, "'Climate Grief': The Growing Emotional Toll of Climate Change," NBC News, December 24, 2018.
9. Duncan McCue, "Growing 'Ecological Grief' Is the Mental Health Cost of Climate Change," CBC Radio, October 21, 2018.
10. Jordan Rosenfeld, "Facing Down 'Environmental Grief'," *Scientific American*, July 21, 2016.
11. The Lancet Countdown on Health and Climate Change, *2018 Report: Briefing for Canadian Policymakers* (2018): 15.
12. Ibid.
13. Joshua Trey Barnett, "Naming, Mourning, and the Work of Earthly Coexistence," *Environmental Communication* 13, no. 3 (2019): 288.
14. Cunsolo and Landman, "Mourn Beyond," 24.
15. Ibid., 16.
16. Renee Lertzman, *Environmental Melancholia: Psychoanalytic Dimensions of Engagement* (New York: Routledge, 2015): 6.
17. Ibid., 4.
18. Timothy Morton, *Dark Ecology: For a Logic of Coexistence* (New York: Columbia University Press, 2014).
19. Nancy Menning, "Environmental Mourning and the Religious Imagination," *Mourning Nature: Hope at the Heart of Ecological Loss and Grief*, eds. Ashlee Cunsolo and Karen Landman (Montreal: McGill-Queen's University Press, 2017): 40.
20. Jie Li, Margaret Stroebe, Cecilia L. W. Chan, and Amy Y. M. Chow, "Guilt in Bereavement: A Review and Conceptual Framework," *Death Studies* 38 (2014): 169.

21. Ibid., 166.
22. Lertzman, *Environmental Melancholia*, 108.
23. Ibid.
24. Derrick Jensen, *The Myth of Human Supremacy* (New York: Seven Stories Press, 2016): 5, 31.
25. Robin Wall Kimmerer, *Braiding Sweetgrass: Indigenous Wisdom, Scientific Knowledge and the Teachings of Plants* (Minneapolis Milkweed Editions, 2012): 55–56.
26. Cunsolo and Landman, "Introduction," 16.
27. Ibid., 6.
28. Richard Gross, *Understanding Grief: An Introduction* (London: Routledge, 2015): 5–6. For additional sources similarly distinguishing grief from mourning, see: John Archer, *The Nature of Grief: The Evolution and Psychology of Reactions to Loss* (London: Routledge, 2003); Therese Rando, "Grief and Mourning: Accommodating to Loss," *Dying: Facing the Facts*, eds. H. Wass and R. Neimeyer (Philadelphia: Taylor & Francis, 1995): 211–241.
29. Ibid., 6, 18.
30. David Konstan, *The Emotions of the Ancient Greeks: Studies in Aristotle and Classical Literature* (Toronto: University of Toronto Press, 2006): 362.
31. Ibid., 244.
32. Beth L. Hewett, "The Eulogy: Grief and the Wisdom of the Ancients," *Sizing Up Rhetoric*, eds. David Zarefsky and Elizabeth Benacka (Long Grove: Waveland Press, 2008): 90.
33. Ibid., 100.
34. Cunsolo and Landman, "Mourn Beyond," 16.
35. Donovan J. Ochs, *Consolatory Rhetoric: Grief, Symbol, and Ritual in the Greco-Roman Empire* (Columbia: University of South Carolina Press, 1993): 25.
36. Susan Clayton, Christie Manning, Kirra Krygsman, and Meighen Speiser, *Mental Health and Our Changing Climate: Impacts, Implications, and Guidance* (Washington, DC: American Psychological Association, 2017): 4.
37. Global Justice Equality Project, "Earth Minute: David Buckel's Self-Immolation in Face of Climate Change," April 27, 2018, https://globaljusticeecology.org/listen-david-buckels-self-immolation-in-face-of-climate-change/.
38. Jeffery C. Mays, "Prominent Lawyer in Fight for Gay Rights Dies After Setting Himself on Fire in Prospect Park," *The New York Times*, April 14, 2018, https://www.nytimes.com/2018/04/14/nyregion/david-buckel-dead-fire.html.
39. Global Justice Equality Project, "David Buckel."

40. Glenn Albrecht, "Solastalgia and the Creation of New Ways of Living," *Nature and Culture: Rebuilding Lost Connections*, eds. Sarah Pilgrim and Jules Pretty (London: Taylor & Francis, 2010): 225.
41. Ibid., 227.
42. Ibid. Etymologically, the first half of *solastalgia* draws on two Latin roots: *solari*, referring to alleviation of distress, and *solus*, which connotes loneliness, isolation, and a general deprivation of comfort. The second half is from *algia*, a New Latin suffix referring to pain and hardship.
43. Glenn Albrecht, "Solastalgia and the New Mourning," *Mourning Nature: Hope at the Heart of Ecological Loss and Grief*, eds. Ashlee Cunsolo and Karen Landman (Montreal: McGill-Queen's University Press, 2017): 299.
44. Albrecht, "New Ways of Living," 217.
45. Ibid.
46. Albrecht, "New Mourning," 297.
47. Ibid.
48. Ibid.
49. Lertzman, *Environmental Melancholia*, 144.
50. Ibid.
51. Ashlee Cunsolo Wilcox, "Climate Change as the Work of Mourning," *Ethics & The Environment* 17, no. 2 (2012): 155.
52. Menning, "Environmental Mourning," 39.
53. Lertzman, *Environmental Melancholia*, 108.
54. Glenn Albrecht, *Earth Emotions: New Words for a New World* (Ithaca: Cornell University Press, 2019). The word guilt only appears twice, neither involving commentary.
55. Robert Macfarlane, *Landmarks* (New York: Penguin Books, 2016): 3.
56. Ibid.
57. Ibid.
58. Ashley Ahearn, "A Seattle Filmmaker Confronts His Grief Over a World That's Changing," Oregon Public Broadcasting, May 2, 2017, https://www.opb.org/news/article/a-seattle-filmmaker-confronts-his-grief-over-a-world-thats-changing-/.
59. Ibid.
60. Cunsolo and Landman, "Mourn Beyond," 6. Emphasis added.
61. Chris Jordan, *Albatross*. Digital film. Dir., Ed., Chris Jordan, Producer, Victoria Sloan Jordan. *Chris Jordan Photographic Arts* (2018).
62. Chris Jordan, "The Story of Albatross," *Albatross*, https://www.albatross-thefilm.com/ourstory/.
63. Jordan, *Albatross*. All additional references to Chris Jordan speaking, unless explicitly cited as otherwise, are quotations from the film.
64. Brooke Jarvis, "The Messengers," *Pacific Standard*, September 8, 2015.
65. Lertzman, *Environmental Melancholia*, 73.
66. Ibid., 6, 73.

CHAPTER 6

Epilogue: The Future of Environmental Guilt

Abstract This epilogue argues that cultivating a new orientation toward environmental guilt can help us engage ecological crisis more effectively. Drawing on Aldo Leopold's land ethic, I propose redefining environmental guilt as *the felt expression of care for biotic communities we have harmed or whose integrity, stability, and beauty we have failed to protect.* When approached as a resource and teacher, environmental guilt can guide us toward ecological connection and care. In order to transform environmental guilt's affective pull toward reconciliation and reparation, however, I contend that we must act as part of something greater than ourselves. Collective guilt calls for collective action.

Keywords Guilt · Hope · Environmental communication · Emotional ecologies · Emotional literacy

"And so we come to the obligatory question about hope," writes environmental philosopher Kathleen Dean Moore in concluding *Great Tide Rising: Towards Clarity and Moral Courage in a Time of Planetary Change*.[1] As the first sentence of the book's final chapter, Moore is winking at readers familiar with the genre's expectations. Turning toward hope when concluding works on ecological crisis is a move so ubiquitous it seems to have shifted from customary to compulsory. Moore engages the issue with characteristic shrewdness and style, delineating

T. Jensen, *Ecologies of Guilt in Environmental Rhetorics*,
Palgrave Studies in Media and Environmental Communication,
https://doi.org/10.1007/978-3-030-05651-3_6

and describing different varieties of hope, providing in the process an exemplary demonstration of emotional literacy.

On the spectrum of hope that she sketches, at one extreme is hope-as-delusion, a lie we receive or give to ourselves to keep moving forward. "False hopes," as Derrick Jensen puts it, "bind us to unlivable situations, and blind us to real possibilities."[2] For Jensen, hope—false or otherwise—is "a longing for a future condition over which you have no agency."[3] Promoted as a "reason for persevering [and] our protection against despair," hope pervades environmental communication and activism.[4] As Moore sees it, operating at the other end of the spectrum is *active hope*. Citing Joanna Macy, Moore describes active hope as a process of "radical imagining," wherein people intentionally and strategically "envision an inspiring future … map out the story of how to get there, and then to identify their role in the story."[5] Far from blind hope or empty optimism, Moore argues that once active hope is integrated into your everyday actions, it transforms into "stubborn, defiant courage [and] principled clarity."[6] At both ends of the spectrum and across the varied assessments of its usefulness, hope is framed as a potent emotional force and the principal motivating trope—for good or ill—of contemporary environmental communication.

Hope has held a monopoly on the emotional ambition of environmental engagement for too long. Its assumed status as superior motivational catalyst for effective action is dubious at best and dangerously oversold at worst. What I'm interested in examining here is not necessarily how or to what extent hope persuades, but rather, the consequences of its preeminent position within the emotional ecologies of environmental communication. In the rush to instill hope, the persuasive dynamics of other emotions have been overshadowed and understudied. As I argued in "Guilty Grief and Ecological Mourning," hope is frequently framed as an antidote to feelings traditionally tagged as having a negative valence, such as grief and guilt. In hastening to evade or sedate these emotions, however, we lose an opportunity to learn from their insights into ecological relation and our feeble collective response to climate crisis and mass extinction. Renee Lertzman takes a similar stance, arguing that in "a climate increasingly fixated on discourses of hope, we tend to avoid difficult emotions," yet in her psychoanalytic approach, "loss and ambivalence [are] psychosocial 'achievements' not to be avoided but integrated for more authentic modes of engagement with a dynamic, uncertain world."[7] As scholars and practitioners of environmental communication, we must expand our emotional aperture, bringing into focus a wider field of feelings that influence our beliefs and behaviors.

In doing so, we can recalibrate our appraisals of emotions. For example, although grief is typically ascribed a negative valence, by reconfiguring it—as Chris Jordan does in the film, *Albatross*—as the felt experience of love for something we have lost or are losing, that negative label becomes less suitable. The same is true for reframing guilt as the felt experience of care for something we have harmed or infringed upon. The approach I'm advocating for would not simply relabel these emotions as having a positive valence, but would involve rethinking the implied linearity of a positive/negative spectrum altogether. An ecological approach to the rhetorical study of emotion eschews such linearity and instead attends to interrelation, circulation, and adaptivity.

To see emotions as ecological in both metaphorical and material senses is to reconsider a host of deep-rooted and detrimental assumptions that structure western cultural discourses and institutions in binaristic frames, especially those assumptions that pit rationality versus emotion, nature versus culture, and human versus nonhuman. Emotional ecologies are more dynamic and intricate than such reductive binaries can account for, though such a claim is hardly revelatory. The entanglement of emotion and rationality has been demonstrated countless times by diverse disciplines. An ever-growing mountain of evidence from neuroscience to rhetorical theory reveals the farce of their contraposition, as well as the inimical ends to which their mobilized. Reason *requires* emotion. "Indeed," cognitive linguist George Lakoff notes, "you cannot be rational without emotions."[8]

Although public discourse is still saturated with manifestations of these entrenched binaries and the assumptions that uphold them, the destabilizing forces of ecological crisis are bringing long-held frameworks into question. We live in a terrifying and wondrous moment, wherein rapid and severe planetary change is occasioning the opportunity to rewrite the axiological narratives by which we live. We are in the midst of a paradigm shift, inescapably tasked with orienting ourselves to volatile ecologies, yet also presented with an extraordinary opportunity to compose new ethics, new aesthetics, and new rhetorics for an animate earth under siege.

Now is the time to identify the affective identifications and emotional frameworks that have facilitated ecological destruction, and simultaneously, to cultivate affective and emotional attunements to the systems that sustain a deep diversity of life. Making sense of our current crisis, in other words, quite literally entails creating new sensorial habits. "To *make sense*," as David Abram puts it, "is to release the body from the constraints imposed by outworn ways of speaking, and hence to renew and rejuvenate one's felt awareness of the world. It is to make the senses wake up to where they are."[9] For such rejuvenated modes to occur at a systematic level, we must forge emotional literacies attuned to and in harmony with ecological systems.

The Unwritten Future of a Keystone Emotion

Given its centrality to how we engage with the destabilization of our climate and our causal role in the sixth mass extinction, we know comparatively little about environmental guilt and its persuasive dynamics. Although this book offers analysis of how it functions as a sensory experience, cultural category, rhetoric of allegation, and catalyst for redressing wrongs, its importance for this moment of crisis demands much more critical attention. The rationale for scholars of environmental communication to be leaders in this regard is straightforward: contemporary environmental rhetorics are fundamentally mediated by issues of culpability, responsibility, and reparation. As a keystone emotion for environmental communication, guilt's principal role in how affective and rhetorical ecologies are organized and navigated merits more thorough accounting.

To move the study of environmental guilt forward, this book examined its rhetorical history and cultural contingency. What is revealed is how our contemporary conceptualization of guilt, grounded in individual culpability and accountability, leads us to overlook a subspecies that operates in relation to widespread injustice or large-scale destruction. Environmental guilt is a form of *collective guilt*, in which the link between individual contribution and overall collective harm may be stretched out conceptually. Phenomenologically, however, it can be felt as intensely as any individualized guilt, if not more so, because the scale of harm is expanded. Our ability to evaluate collective guilt's rhetorical register in the scenarios of scale through which it circulates, however, is inhibited by the prevailing notion of guilt, which is heavily premised on the individual citizen and their intentional actions.

In "Guilt's Plasticity" and "Eco-friendly Scapegoats," I pursue the implications that flow from this relatively recent configuration of guilt, chief among them are impoverished emotional literacies for grappling with collective guilt. We struggle to identify, contemplate, navigate, and communicate feelings of environmental guilt, which leads to confusion, isolation, and vulnerability. Ultimately, without frameworks attuned to its unique dynamics, feelings of collective guilt can be exploited, rendering it a powerfully individualizing force that attenuates possibilities for collective action. The more we understand about environmental guilt—how its affective patterns influence our attitudes and actions, how it's provoked and perpetuated—the better we'll understand environmental communication.

The analyses provided in the previous five chapters aim to lay a foundation for the future study of environmental guilt. Recognizing that guilt has a past implies that it has a future—ideally, one in which we are more attuned to the ecologies that sustain us and others of all species. In this epilogue, then, I suggest some tools and theoretical frameworks that will perhaps prove productive for the challenges we face in the uncertain times before us.

Transforming Environmental Guilt: Next Steps

In his Afterword to *Tracing Rhetoric and Material Life: Ecological Approaches*, Nathan Stormer writes, "[e]cological rhetoric names an intensification of rhetoric becoming ecological."[10] Citing Jenny Rice's "watershed article," "Unframing Models of Public Distribution: From Rhetorical Situation to Rhetorical Ecologies," Stormer observes that distinctions remain, albeit subtle ones, between "human, social environments" and nonhuman natural environments in her description of affective ecologies.[11] "This is ecology becoming rhetorical" he remarks, signaling that Rice's article helped tilt the field's theoretical gaze toward rhetoric's ecological dimensions, yet leaves ample room for advancing further.[12]

Stormer's analysis is not a critique of Rice, however, as much as it is a time-stamp, a marker of where the field stood in relation to ecology in 2005. Some decade and a half later, considerable maturation in theorizing both rhetorical ecologies and ecological rhetorics can be seen. The latter is evident in Caroline Gottshalk Druschke's and Candice Rai's essay, "Making Worlds with Cyborg Fish," wherein "rhetoric here is experienced as ecology for ecology. Or, instead of conceiving rhetoric as epistemic, it is *feeling rhetoric* as *ecosystemic*."[13] The vitality and urgency surrounding rhetoric's engagement with ecology has grown and is poised to intensify. Rhetorical theory is undergoing an ecological turn, reformulating foundational concepts and frameworks in the process.

The stock-taking above is to acknowledge how the field has progressed and where it is headed. The rhetorical study of environmental guilt aligns with its current trajectory and theoretical engagement with affect, materiality, and addressivity. In many ways, the next steps in transforming our understanding of environmental guilt is for rhetoricians to keep proceeding in the direction we're headed with our broader theoretical inquiries. As we do so, we would be wise to enrich our vocabularies for emotional ecologies, establish emotional literacies as a standard facet of rhetorical pedagogies, and, finally, develop habits of framing guilt and other emotions as teachers with the capacity to instruct us and to help us cultivate ecological attunement.

"We know enough of our own history by now to be aware that people *exploit* what they have merely concluded to be of value," Wendell Berry declares, "but they *defend* what they love. To defend what we love we need a particularizing language, for what we love we particularly know."[14] The shallow vocabulary we currently employ for guilt is inadequate and our deficient frameworks contribute to the emotional disorientation that facilitates ecological exploitation. We need a particularizing language for the study and everyday navigation of emotional ecologies. This work is not merely analogous to the duty Berry describes; it is fundamentally a part of that effort. Massive ecological devastation would not occur with such systematic efficiency and so little revolt without equally massive emotional disorientation. Enriching our attunement to emotional orientations and affective attachments—especially those that facilitate exploitation and those that effectively resist it—is critical work, not as an end in itself, but so that we can be more effective in reversing ecocide.

The value and impact of this work will be severely limited if it does not also become integrated into classroom pedagogies that invite students across all levels to participate in refashioning emotional literacies. Teachers of environmental communication, and rhetoric more generally, are well-positioned to reintroduce emotional literacy as a customary component of rhetorical education. As heuristics employed to distinguish feelings, detect contributing factors, and evaluate how they orient us toward some actions and away from others, emotional literacies are a pragmatic venue for pedagogical engagement on environmental issues.

This task will not be easy. As I have argued elsewhere, education in *pathos* has hardly changed over the past half-century, despite the fact that scholarly understandings of affect and emotion have advanced in both scope and sophistication.[15] For interested teachers, I recommend a frank conversation with students about the fact they've been unwisely cautioned over emotion's rhetorical force. The next step is to replace the warnings and constraints they've previously received with curiosity and complexity. The analytical exploration of emotion is apt to be a novel experience for students, provoking both challenges and exhilaration. In my experience, students are often eager to inquire into the persuasive energies of emotion, particularly in the context of ecological crisis. This is especially true for discussions of environmental guilt, which can generate lively conversation and, moreover, frequently yield moments of collective catharsis. As this book has shown, we are ill-equipped to confidently navigate the intricacies of environmental guilt with existing frameworks. Providing a space for students to share their experiences and pose questions is an important step in transforming our understanding of this keystone emotion.

The recommendations I've provided thus far—to enrich our lexicon of emotional ecologies and to reintroduce emotional literacy as an explicit component in our curricula—involve, admittedly, little risk. They do entail significant labor and are worthwhile endeavors that will yield benefits for those who undertake them. But with all the world at stake, isn't advocating for more and better words and classroom discussions comically tame? I sympathize with the hypothetical critique, though I would ultimately disagree and reassert their importance. Affect and emotion are the threads that bind us within tapestries of social and ecological relation. As we undertake the effort to create a more livable future for ourselves and our nonhuman kin, having more people recognize emotional textures, histories, power, and potential is hardly ineffectual. In bringing this study to a close, then, I offer one last recommendation that may seem equally tempered, yet if genuinely engaged, contains radical possibilities: approach environmental guilt as a teacher who can guide us in cultivating ecological attunement.

On the surface, such an ostensibly bland proposal hardly seems capable of producing profound transformations that impact everyday behavior. Allowing environmental guilt to direct you to the boundary lines of transgression for which you feel guilt or were accused of being guilty, however, can catalyze a process of inquiry and illumination that yields acute consequences. For guilt to become a teacher—and not a foe to be vanquished—it must be reframed. Mirroring Chris Jordan's reconfiguration of grief, I've proposed guilt be understood as the felt expression of care for something we have harmed. We can further develop this reframing by drawing on Aldo Leopold's formulation of the ***land ethic***, in which "the land" is shorthand for the collection of biotic communities that include soil, water, flora, and fauna. Leopold defines this ethical orientation thusly: "A thing is right when it tends to preserve the integrity, stability, and beauty of the biotic community. It is wrong when it tends otherwise."[16] From this principled framework, we may advantageously define environmental guilt as ***the felt expression of care for biotic communities we have harmed or whose integrity, stability, and beauty we have failed to protect.*** Just as grief's poignant feelings of loss may be elucidated as the felt experience of love, so too can environmental guilt be radically reorganized in a manner that orients us toward care and ecological connection.

Under this revised understanding of environmental guilt, our affective energies move us in a different direction. Distress over the disintegration of biotic communities and stinging contrition for being complicit in that destabilization become receptive to transformation, turning into powerful sources for revealing and reconnecting with the ecological communities we may have harmed. When understood as a teacher, these affective forces of guilt can illuminate the boundaries transgressed, inviting us to consider more deeply the ethical, social, and rhetorical systems that create and constellate those boundaries. A sense of care is essential, after all, to animate guilty feelings. Even when care manifests in the uneasy resonances associated with guilt, we can use that experience to recognize, gain focus on, and learn to act on behalf of the ecological connections that matter.

When we release our resistance to environmental guilt, we allow the complex of sensations and thought patterns to flow where they may. We open ourselves to a deeper understanding of our guilt and we center our attention on the ecological communities in which we are enmeshed, and whose integrity, stability, and beauty we care for. Moreover, by not rushing to evade or extinguish guilt, we can become more attuned to our role in the harm and more keenly delineate the dimensions and extent of our responsibility to redress that harm. In this process, we may clarify for ourselves how feelings of ecological care and connection are, today, often strategically isolated and misdirected to perpetuate an unjust and inherently unsustainable socioeconomic system. Herein lies the latent volatility. *Feelings of environmental guilt are recognitions of connection to ecological systems that precede and exceed specific cultural formations.* As such, their educative potential is a means of ecological attunement and a source for generating creative, reparative action.

At the moment, feelings of collective guilt over ecological harm too often become an individualizing force. They need not be in the future. The desire for making amends that accompanies collective guilt cannot be satiated through individual actions. To transform environmental guilt's affective pull toward reconciliation and reparation, we must act as part of something greater than ourselves. Collective guilt calls for collective action.

Notes

1. Kathleen Dean Moore, *Great Tide Rising: Towards Clarity and Moral Courage in a Time of Planetary Change* (Berkeley: Counterpoint Press, 2016): 311.
2. Derrick Jensen, "Beyond Hope," *Orion* 25, no. 3 (2006): n.p., https://orionmagazine.org/article/beyond-hope/.
3. Ibid.
4. Ibid.
5. Moore, *Great Tide Rising*, 315.
6. Ibid., 318.
7. Renee Lertzman, *Environmental Melancholia: Psychoanalytic Dimensions of Engagement* (New York: Routledge, 2015): 23.
8. George Lakoff, "Why It Matters How We Frame the Environment," *Environmental Communication* 4, no. 1 (2010): 72.
9. David Abram, *The Spell of the Sensuous: Perception and Language in a More-Than-Human World* (New York: Penguin Random House, 1997): 265.
10. Nathan Stormer, "Afterword: Working in an Ecotone," *Tracing Rhetoric and Material Life: Ecological Approaches*, eds. Bridie McGreavy, Justine Wells, George F. McHendry, Jr., and Samantha Senda-Cook (New York: Palgrave, 2018): 345.
11. Ibid., 345.
12. Ibid., 345.
13. Ibid., 349. Emphasis added.
14. Wendell Berry, *Life Is a Miracle: An Essay Against Modern Superstition* (Berkeley: Counterpoint Press, 2001): 86.

15. Tim Jensen, "Textbook Pathos: Tracing a Through-Line of Emotion in Composition Textbooks," *Composition Forum* 34 (2016).
16. Aldo Leopold, *A Sand County Almanac: And Sketches Here and There* (Oxford: Oxford University Press, 1949): 204, 224.

Index

T. Jensen, *Ecologies of Guilt in Environmental Rhetorics*, Palgrave Studies in Media and Environmental Communication, https://doi.org/10.1007/978-3-030-05651-3

The manufacturer's authorised representative in the EU is Springer Nature Customer Service Centre GmbH, Europaplatz 3, 69115 Heidelberg, Germany. If you have any concerns regarding our products, please contact ProductSafety@springernature.com

Printed and bound by CPI Group (UK) Ltd, Croydon, CR0 4YY
30/09/2026
02225472-0012